CONQUERING MATH ANXIETY

A SELF-HELP WORKBOOK

Cynthia A. Arem, Ph.D.
Pima Community College

Brooks/Cole Publishing Company
Pacific Grove, California

TO ARNIE

My best friend, confidant, teacher, guardian angel, and nurturer of all my dreams and ambitions—so loving, sensitive, and sincere. Thank you for being there, every step of the way in my life and in my work.

Consulting Editor: Karl J. Smith

Brooks/Cole Publishing Company
A Division of Wadsworth, Inc.

Printed in the United States of America

10 9 8 7 6 5 4 3 2

Library of Congress Cataloging in Publication Data

Arem, Cynthia
Conquering math anxiety : a self-help workbook / Cynthia Arem.
p. cm.
Includes bibliographical references
ISBN: 0-534-18876-1
1. Math anxiety. 2. Mathematics--Study and teaching. 3. Self-help techniques. I. Title.
QA11.A77 1992
370.15'651--dc20 92-18880
CIP

Sponsoring Editor: *Paula-Christy Heighton*
Editorial Assistant: *Carol Ann Benedict*
Production Editor: *Linda Loba*
Manuscript Editor: *Barbara Kimmel*
Interior Design: *Jill Wood*
Cover Design: *E. Kelly Shoemaker*
Art Coordinator: *Lisa Torri*
Interior Illustration: *Ryan Cooper*
Cover Photo: © *The Stock Market / Paul Steel*
Typesetting: *Patricia Douglass*
Printing and Binding: *Malloy Lithographing, Inc.*

PREFACE

Millions of students in this country are terrified of math. They do whatever they can to avoid numbers and math problems as if they were the plague. In school, they put off taking math as long as possible. And, when given the choice, they select majors requiring little or no math. If they do have to take the subject, they dread entering the classroom, let alone taking an exam. One young co-ed described how she panicked in math class to the point that she would run out of the room and vomit uncontrollably. As a child, she often had nightmares about numbers chasing her, wanting to hurt her.

High levels of anxiety can devastate a student's ability to perform, resulting in poor academic progress and high dropout rates. Research among college students has shown that the activity of math itself generates anxiety reactions among students who are not necessarily highly anxious in other situations. One study at a large midwestern university disclosed that one-third of all students who requested behavioral counseling complained of anxiety related to math (Suinn, 1972).

As a college counselor, I see hundreds of students each semester who fear math or do their best to avoid taking it. In fact, I believe that math isn't a problem for students only; there are people in all walks of life who are math anxious. Invariably, when I'm at social functions or at meetings, people confide in me about their math woes. They often tell me that math had always been their biggest problem in school. They say things such as: "If it weren't for math, I'd be going to school right now." "I'd be making a career change." "I'd be more successful than I am right now." "I'd go back to school and get that degree I always wanted." And on and on.

Widely reported studies have shown that math serves as a "critical filter" in determining many educational, vocational, and professional options (Sells, 1978; National Research Council, 1989). Math avoiders are finding themselves shut out of many of today's most rewarding and profitable careers.

College algebra is a minimum math requirement for the bachelor's degree at almost all undergraduate institutions. At the community college level, we see many students who can't possibly meet this math requirement, either because they had poor math backgrounds to begin with or because they've been out of school so long that they've forgotten what they once knew. Often these students must take four math courses just to meet this university requirement for their major. They need to take fundamentals of math—a course that reviews percentages, decimals, and fractions—before they can go on to take beginning algebra, intermediate algebra, and finally college algebra. For students who are afraid of math, this is a living nightmare. Furthermore, the news media increasingly point out that the majority of Ph.D. degrees in science and engineering given in this country are now being awarded to foreign-born graduates who, presumably, are more comfortable with math than U.S. graduates.

Why is math such a problem for Americans? How can we fully understand this problem? What can be done to overcome the devastating effects of math anxiety and math avoidance? I've designed this workbook to answer these questions.

Conquering Math Anxiety: A Self-Help Workbook presents a comprehensive, multifaceted treatment approach to math-anxiety reduction and math avoidance. This unique, carefully outlined methodology involves anxiety management and reduction, reduction of internal psychological stumbling blocks, attitude changing, confidence building, "Success in Math" visualizations, learning-style enhancement, and effective study and test-taking skills. In addition, it teaches winning strategies for overcoming test anxiety on math exams. This workbook provides detailed explanations along with a host of varied exercises, methods, and worksheets for helping math-anxious students deal with and overcome math fears.

Although *Conquering Math Anxiety* is based on my experience at the community college level, my primary purpose in writing it is to reach *all* math-anxious individuals. A great many of these people are attending universities, community colleges, or high schools. Many are no longer in school but find that math fears continue to interfere with the pursuit of their personal goals, often creating lowered self-esteem, frustration, or dissatisfaction.

Because this workbook is comprehensive in its scope, it is relevant for people at any educational level or in any career. It is designed so that the reader can simply jump in and draw from it as little or as much as he or she needs.

No prerequisites are needed for completing the activities in this workbook. Mathematics anxiety can be found at any math level, from the fundamentals of math through statistics and beyond. I've even had students taking calculus simultaneously enrolled in my math-anxiety reduction course.

Conquering Math Anxiety uses the analogy of a road map detailing the route to math success, but its individual chapters are like recipes that can be extracted and used entirely on their own according to need. For this reason, the workbook is useful by itself or as a supplement to any mathematics course in which a student or teacher identifies a problem area that impedes smooth progress.

My background as a college counselor, teacher, and workshop trainer leads me to speak directly to those with whom I work and to involve them thoroughly in the learning process. Therefore, I have included a maximum number of exercises. Also, throughout the book, I address the reader in the first person. After first analyzing the roots of math anxiety, I detail many specific techniques for managing anxiety and overcoming psychological barriers to math. This leads to the critical importance of reversing negative "math self-talk," creating positive attitudes, and building self-confidence. The section on math success visualization is unique. It helps the readers to reprogram their thinking in a way that will promote math success. Then, I focus on learning styles and study skills as they apply specifically to math. The workbook ends with an in-depth look into techniques for conquering math test anxiety and an analysis of the importance of mathematics in all aspects of life. Interspersed throughout the workbook you will find humorous cartoons with a positive slant.

ACKNOWLEDGMENTS

I am extremely grateful to those who have assisted me in the development of the ideas presented in this workbook: to the many math-anxious students I have counseled or taught; to my colleagues who shared ideas and teaching suggestions; and to my family and friends for their support and assistance.

This work would not have been possible if not for the many wonderful students I have worked with over the years—both in counseling sessions and in the math-anxiety reduction courses and workshops I have taught. In the process of helping them deal with math anxiety, I have learned much and now feel privileged to be able to share this knowledge with others. My students have been my teachers, and I would like to thank each and every one of them.

The most important person in the development and writing of this book is my husband, Arnold Arem, M.D. His many hours of dedication to this project have been invaluable to me. To each of the ideas presented here he listened carefully and gave his thoughtful critique. He read and reread my manuscript at each stage of its development and offered me counsel, detailed editing advice, and much needed emotional support. I am particularly delighted that he developed the ideas for the lighthearted cartoons that appear in this work. I would also like to thank my daughter, Kim, for her editorial comments on my manuscript.

A wonderful colleague and friend I want to thank is Debbie Yoklic. Our association has been a rewarding one of shared ideas, team-teaching strategies, research efforts, and curriculum development for a course on math anxiety reduction at our college. Our contacts over the years and experience with our course were a stimulus for the writing of this book. I also would like to thank her for her valuable suggestions for improving this book.

I would like to thank my good friend and colleague Arlene Scadron for reading portions of my manuscript and offering editorial advice.

I am grateful, too, to all my colleagues in the Mathematics Department of Pima Community College; in particular I want to thank Lou Ann Pate and Mehdi Sadatmousavi, who have taught the math anxiety reduction course with me, and Charles Land and Sam Borah who offered me many thoughtful suggestions. I want to thank Frank Rizutto for his support and Gary Mechler for the stimulating talks we've had concerning Chapter 10. Others at Pima College who have been very supportive of my ongoing work in math-anxiety reduction and whom I would like to acknowledge with gratitude are: Ken McCollester, the Associate Dean of Math and Sciences; and Michael Curry, Acting Associate Dean of Business, Computer, and Human Sciences.

I would like also to thank the following reviewers who read my manuscript and offered excellent editorial comments: Professor Josette Ahlering, Central Missouri University; Professor Denise Brown, Collin County College; Professor Virginia Carson, DeKalb College; Dr. Robert Main, Western Oregon State College; Professor Carol Nessmith, Georgia Southern University; Professor Richard H. Schwartz, College of Staten Island; and Dr. Karl Smith, Santa Rosa Junior College.

The staff at Brooks/Cole also deserve special credit. In particular, I want to thank Paula Heighton for her great ideas, market research, continued encouragement, positive feedback, and delightful conversation. Her excellent guidance throughout this project was invaluable. My gratitude also to Carol Benedict, who was always supportive and ready to assist me when I needed help. My special thanks to Linda Loba, Production Editor, who made me an integral part of the production and design process, and whose clarity of vision created an outstanding finished product. To Lisa Torri, Senior Art Coordinator, Carolyn J. Crockett, Advertising Coordinator, and the myriad wonderful people behind the scenes who made it all happen so smoothly and pleasantly, my deepest thanks.

Cynthia A. Arem

CONTENTS

INTRODUCTION: THE ROAD MAP TO SUCCESS

Have you ever found yourself lost, in a strange or unfamiliar place, without a map to guide you? Did you suddenly feel alone? Did you feel scared, not knowing what to do or which way to turn? Perhaps you stopped to ask for directions, but they were confusing, misguiding, or incomprehensible. If the hour was growing late, you might have felt a sense of menace or dread. Perhaps you began to perspire; your muscles tightened, your pulse quickened, your breathing became more rapid, and all your senses seemed heightened, trying to detect if any danger were present.

Now let's imagine for a moment that you were miraculously given a detailed map of the area, not only showing where you were, but also providing explicit instructions for reaching any destination you chose. Every step along the way was made clear, every turn and every landmark carefully spelled out and described. Your map left no stone unturned, no shadow of a doubt as to what direction to take and how to proceed. Wouldn't such a map have made your journey less stressful, more pleasurable, and helped guarantee success in reaching your goal?

Does dealing with math seem like being lost in an unfamiliar or strange place? Do you sometimes feel as if you're all alone and that no one quite understands what you're going through or experiencing? And, even when others try to help, it just seems confusing or misleading to you? When doing math, do you ever feel scared, panicky, or have a sense of dread or impending doom—not knowing which way to turn or where to get the assistance you so desperately need?

Let me assure you that you are not alone. There are many people who have felt the same way at one time or another. Students often describe feelings of devastating failure and utter defeat when it comes to doing math. Mathematical reasoning and numbers are their nemesis and have been for much of their school life. They shun exposure to math as if it were a frightening, unknown danger. Whenever possible, they purposely avoid any subject requiring math and choose majors requiring little, or preferably, no math. In math classes, or even in math tutoring sessions, their minds seem to turn off and they become easily lost, comprehending little of what is explained or taught.

If you are one of these students—lost, feeling alone, apprehensive, shunning math—you have come to the right place. I have written this workbook specifically for you. I've designed it to be your personal, detailed map to help you find your way, to aid you in overcoming your fear of math, and to ensure that you achieve success in math. Included are detailed explanations and exercises all along your path to illuminate your journey and to make it easier and more fulfilling.

As you follow this road map, you will find many important branch points for you to explore and experience. We begin our journey in Chapter 1 by first discovering whether you really have math anxiety, and then we analyze your math success goal. From there, we'll look at where your math fears and anxieties initially began and why they have persisted.

Chapter 2 asks you to look at your math anxiety history, to identify the factors that influenced you and the myths, stereotypes, or games that might have affected you. By diagramming the math anxiety process, you can see how previous negative math experiences combined with negative self-talk lead to overwhelming anxieties and fears.

In Chapter 3, you learn how to control and manage those anxieties so that they can work for you and not against you. You learn how extremely high, uncontrolled anxiety levels lead to panic and the inability to perform, whereas moderate, properly managed anxiety marshals optimal performance, good memory, and clear thinking.

Chapter 4 explores how you deal with problems that arise in your life and how this can aid or hinder your efforts to overcome math anxiety. The "Wall Fantasy" is specifically designed to increase your awareness of what you may be doing to impede your success in math.

Next, in Chapter 5, we look at the importance of attitudes and ways of changing negative attitudes toward math to positive ones. I guide you through exercises to help you develop positive, enhancing self-dialogue related to math and your ability to do math.

Then our journey takes us to the very exciting area of "Success in Math" imagery. Research has shown that positive mental imagery can greatly improve the status of our lives mentally, emotionally, and physically. In Chapter 6, you are given excellent visualization techniques to achieve your math goals. You are guided through a beautiful-imagery technique that has already advanced the lives of many hundreds of math students who overcame the same math anxieties that you have been experiencing.

Chapter 7 looks at ways and means of enhancing your learning. Through the use of carefully chosen questions, you are asked to assess the critical factors that positively influence your ability to learn math. You'll be surprised at how the right formula for combining these factors can help you learn more efficiently.

From here, the path leads directly into Chapter 8, where we examine useful math study skills and winning strategies for learning math and reading math textbooks.

Chapter 9, one of the most important in this road map, teaches you how to deal with test anxiety and how to perform at your best on math exams. It will help you learn to feel alert, clear, calm, confident, and competent on exams. It reviews test preparation and test-taking strategies, as well as good nutritional guidelines to follow before exams. We end the journey with Chapter 10, a look at your exciting math future.

SOME BASIC AXIOMS

In this workbook, I present a comprehensive, well-tested plan for tackling your math fears and anxieties. My approach is based on the following premises:

- As a student, you need not be forever burdened by the negative experiences or unproductive messages from your past.
- You can learn to manage stress and anxiety physiologically so they can be productive rather than destructive aspects of your academic performance.
- By changing negative math self-talk to positive self-talk, you can greatly improve your ability to deal with math.
- Learning how to maintain a positive attitude toward math and toward your ability to do math has a ten-fold beneficial effect on your math performance.
- By using visualization techniques, you can reprogram yourself to succeed in math.
- You can increase your ability to learn math by using more appropriate, effective, and efficient study skills and learning-style strategies.
- Conquering math test anxiety is like winning the battle.
- Math can actually begin to be fun and exciting for you!

GUIDELINES FOR USING THIS WORKBOOK

Here are a few suggestions that may help you maximize the benefits of the exercises in this workbook:

1. *Give yourself some uninterrupted time.* Set aside a special time and place where you will not be distracted or interrupted. Even if you have only five or ten minutes, make sure it is a private time just for you. Unplug the phone if necessary. Put a note on your bedroom door saying you can't be disturbed for a few minutes.
2. *Work alone and in silence.* You will be able to gain greater self-understanding and insights if you work alone and don't prematurely share your new perceptions with others. There will be plenty of time later, after completing your work, for you to discuss what you've learned. I encourage you to share your insights with supportive people, such as friends, teachers, or a counselor. They may make your journey easier.
3. *Keep a math journal.* You will probably find it helpful to keep a journal in which you write about the new awareness, sensitivities,

and insights you gain. This may be similar to Exercise 2-3 (Chapter 2), or it may be in the form of a traditional diary.

4. *Assume a comfortable position.* Sit quietly, with your eyes closed, in a slightly darkened room when doing: the anxiety management exercises in Chapter 3, the Wall Fantasy in Chapter 4, the "Success in Math" visualization in Chapter 5 and the "Math Test Anxiety Reduction Visualization" exercise in Chapter 9. I suggest that you record the instructions for these exercises onto an audiocassette, so that you can listen to them repeatedly whenever you wish. Be sure to arrange to be alone and undisturbed so you attain the full benefit of these experiences.

5. *Be persistent.* The road to success can only be achieved through staying power, resolve, and determination. "Stick-to-it-tiveness" will help you reach your math goal. You must master many steppingstones along the way: from learning how to reduce anxiety to overcoming psychological stumbling blocks; from rewriting disempowering math beliefs to reprogramming yourself to succeed in math; from learning to use effective math study skills to conquering math test anxiety. It will take time and patience. So persevere, stay with it, and you'll conquer your math fears and anxieties.

6. *Be positive.* Know in your heart and mind that **you can and will suceed.** Have faith that you are capable of achieving in life what you realistically desire and work toward. Your persistent, positive efforts can and will be rewarded. I've seen it happen with others who have conquered math anxiety. Why not you?

7. *Jump in and start anywhere.* Although I have written this workbook as a detailed road map to help you reach your math success goal, you need not progress through the workbook in sequence. As much as possible, each chapter stands independent of the others. Thus, you may wish to work mainly on those areas that concern you the most.

SUMMARY

Conquering Math Anxiety: A Self-Help Workbook offers you a detailed road map laying out all the steps along the route to overcome math anxiety and ultimately to achieve success in math. Many students have taken this route before you and have succeeded. So can you!

CHAPTER

1

DO YOU REALLY HAVE MATH ANXIETY?

Knowing what kind of aid and direction you'll need depends first on where you are now and where you want to go. I've found that much of the anxious, blocking, fear-stricken behavior that many students experience in math often isn't primary to the subject of math but is situational. In other words, anxiety appears to arise in math class, but it may not be caused by math itself. It seems that math acts as a fine magnifying lens, bringing into sharp focus a host of other academic deficiencies such as poor study skills, knowledge gaps, or inadequate test preparation or test-taking skills. Anxiety expressed about math thus becomes a symptom and not the disease itself. Labeling yourself "math anxious" is almost like having "Medical Student's Disease." Commonly, medical students learn the symptoms of an ailment and then, if they themselves manifest even one of the symptoms of that disease, they are sure they must actually have the disease themselves.

Not unlike a disease, math anxiety (as I define it) is a clear-cut, negative, mental, emotional, and/or physical reaction to mathematical thought processes and problem solving. It is often caused by negative experiences with math in childhood or early adolescence.

Often, students who believe they are math anxious in reality are merely victims of test anxiety. Jonathan, an 18-year-old, is a perfect example of this. He sought counseling from me, distraught because he was failing math but felt he really knew the material. He was sure his poor performance was due to a math block or perhaps a deep-seated

fear of math. Upon questioning Jonathan, it was obvious he did know his math. But when he went into the exam, he would totally "blank out." Sometimes it was as if he couldn't even add two numbers together without making a mistake. I learned that he occasionally blanked out or panicked on exams in other subjects as well. Halfway through a chemistry exam, he panicked and began making all sorts of errors on problems he really knew how to answer. He also found himself anxious and perspiring a lot during his psychology final and ended up ruining his chances for an A in the course.

If you are like Jonathan, panicking on math tests as well as on exams in other subjects, the help you need is primarily in the area of overcoming test anxiety. A reduction in the high levels of generalized anxiety you experience when taking tests will increase your ability to perform, not only in math but in all your courses. Chapter 9 in this workbook, entitled "Conquering Test Anxiety," is an excellent chapter to study. It will give you tried-and-true strategies to prepare for tests, take tests, deal with the anxiety, and use visualization techniques to ensure successful results.

I have also worked with other students who were not truly math anxious, but whose academic difficulties severely affected their ability to succeed in math. Perhaps you are one of these students. For example, you could easily find yourself feeling overwhelmed, anxious and "over your head" in a math course if, unknowingly, you had missed important preparatory course work along the way. You may be surprised to learn that your anxiety in math might abruptly end if only you were properly placed in an appropriate math class or were to receive remedial help in your deficient areas. I have worked with many students who were experiencing severe anxieties in college algebra because of sizeable knowledge gaps in their math background, despite the fact that they had never skipped a semester of beginning or intermediate algebra. Upon further examination, I learned that, although these students had never actually missed any classes, they were struggling with basic algebraic concepts like graphing a line or solving fractional and quadratic equations. For these students, it was like taking German 4 when they had missed learning how to conjugate in German 2. They were simply ill prepared for their current course. They have good reason to be anxious.

If you have missed information along the way, it's almost impossible to expect to understand clearly any subject. In learning a foreign language, the early courses are designed to provide the grammar and syntax for later courses. Each course builds upon the previous ones. Early math courses also provide the "grammar and syntax" of what is to come later—that is, it's the "language" of math. In math, in particular, each step is an essential stepping-stone for the next. When constructing a large, impressive skyscraper we must be sure that the foundation is strong and sturdy. And as we go higher and higher in math, it is like building a tall, elegant, though intricate structure. If you have missed important knowledge along the way, you may have to go back, find those missing blocks, and reconstruct your building. Once you do this and your building sits on a solid foundation, you'll find that math really can become both rewarding and fun.

Fine, you say, but how do you accomplish this? There are a number of options open to you. If the knowledge you are missing is substantial, you should consider actually repeating or, if possible, auditing some of the math courses you have already taken in the past. Many colleges offer developmental math courses, such as the fundamentals of math and beginning algebra. Also, often available are courses that are modularized or separated into smaller units in which you can elect to take only the necessary unit(s) to fill in your background. I've met students who were deficient in some specific information taught in the latter part of Algebra 1 who were delighted to learn that they were able to register for just the third module of an Algebra 1 course at our college.

Ellen is another student who wasn't truly math anxious but whose academic difficulties affected her ability to do math. Ellen understood her math when it was presented in class, but when she sat down to do her homework two days later, it reminded her of hieroglyphics, mysterious and undecipherable. She fell behind in her class work, and soon she became distraught with math. At this point, Ellen came to me, convinced that she was math anxious. When she described to me how she studied math, that she failed to review her math immediately after class and didn't make up her own practice exams (among other things), I was sure Ellen's problems were related to poor study and test preparation skills and not to math anxiety. I spent several sessions with Ellen, teaching her good study strategies. Soon Ellen's grades—and her comfort level—began to improve. If your discomfort in math is related to poor study skills or inadequate test preparation, Chapters 8 and 9 in this book can help you.

GOALS FOR SUCCESS

Before we continue, let's be very clear about where you are going. What are your goals for success in math? What do you want to achieve? If you're taking a math class now, what do you want to get out of it? Why are you taking the course? Why did your school think this course was important in your curriculum? What grade is acceptable to you? What does successful accomplishment in math mean to you? Goals are like anchors to the future we desire. We must toss them out in front of us and then use them to pull ourselves along. In this way, you take control of your math future. Your math success will happen by your design and not by chance. You *can* have the math success you want!

In Exercise 1-1, I'd like you to identify what it is you wish to achieve in math. Write down this goal. Be specific and put it in measurable terms. For example, you might state: "I want to successfully complete Calculus 1," or "I want to pass my statistics course," or "I want to solve quadratic equations," or "I want to be able to work out the word problems in my textbook." It's important above all that you are clear that this is the goal you wish to achieve, and that it's presented without any alternatives. The less conflicted you are about achieving your goal, the greater the probability of accomplishing it.

Next, put down a target date for reaching your math success goal. This is the date when you think you realistically can achieve your goal. Some students find that it's best to set three dates, one being the most optimistic date, one being a more moderate and realistic date, and the third being the latest acceptable date.

The third part of the exercise asks you to evaluate the strength of your goal, to help you to determine how strongly you are motivated to achieve it and what its value is to you. Check off all statements that seem to fit you.

In the fourth part of this plan, establish the benefits you will get from reaching your goal. List both the tangible as well as intangible benefits; sometimes the intangible ones are more important than the tangible ones. For example, Theresa wrote that succeeding in reaching her math goal would make her feel better about herself and help her to know that, one day, she could be successful as an architect. For Carl, succeeding in math would prove that his seventh grade teacher was totally wrong when he said Carl would never be able to do well in math.

In the fifth part of the exercise, identify some of the obstacles you may have to face along your path, as well as steps you may be able to take to overcome these obstacles. Andrew, a college freshman, identified one of the major barriers to his success in math: poor study skills. He decided to research the best study strategies for improving his math and to take a special math study skills workshop offered at his college. Emily's major obstacle was her own negative and embarrassing past history with math, which seemed to loom over her like a large, dark rain cloud. She knew this stopped her from progressing and that she had to take some drastic steps to deal with it. Emily decided she needed to seek counseling to work on her poor self-esteem.

In part six of the exercise, after looking at the barriers and the possible steps required to overcome these barriers, try to discern the positive forces and abilities you can use or strengthen in order to meet your goal. Lucy noted in her plan that she has lots of determination, she is motivated, and she is willing to work hard. Harry felt that his good study habits and ability to overcome obstacles were his biggest assets.

Next, specify the supportive people in your life who can help you on the road to success in math. Who are these people, and how will they be able to help? Perhaps you have a teacher who is very encouraging, or a friend who is a great math tutor, or a companion who is particularly reassuring.

In the eighth section of this plan, I ask you to select the significant action steps you are willing to take in order to meet your math success goal. Include here the measures you think will work best. For example, you may explore the availability of math tutorial programs or decide to work on a specific number of math problems at each study session. Many of these steps you will be aware of only as you continue to read further in this book. As you read, be sure to occasionally return to this list and add relevant steps to it, thereby further personalizing this plan for you. For each action step, list a specific target date by which you will accomplish it.

The last section of this plan for math success asks you to determine how you will reward yourself for meeting your goal. The reward is a very important part of this plan. It will help keep your motivation high and sustain your efforts along the path to success. I've seen all types of rewards described on finished plans, from a night out on the town to a trip to a favorite vacation spot to buying a special gift for yourself. Remember that giving yourself a reward at the successful accomplishment of your goal is a very important part of this plan.

EXERCISE 1-1: MY PLAN FOR MATH SUCCESS

1. I list one realistic math success goal I wish to achieve. I state it in specific, positive, measurable terms. I write out, vividly and in detail, exactly what I want. My math goal is:

____

____

____

____

2. The realistic target date for achieving this goal is:

3. My math goal should meet the following goal-setting criteria (check all that apply):

   ______ I have clearly stated it.

   ______ I value it.

   ______ I believe I can do it.

   ______ I want to do it, and I am motivated.

   ______ I find it rewarding and personally fulfilling.

   ______ I am clear that this is what I want, as opposed to other choices.

   ______ It is a realistic possibility for me in terms of my time and ability.

   ______ I envision a plan of action for achieving it.

4. I want to achieve my math goal because of the following benefits and potential satisfactions (List as many as possible; include both tangible and intangible benefits):

5. These are some barriers or obstacles I may face and steps I will take to overcome them.

Barriers	**Steps to Overcome Barriers**

6. These are the positive forces and abilities I can use or strengthen to meet my math goal:

7. These are the people who can help me in achieving my goal:

Name	**Type of Help They Can Give**

8. The significant *action steps* I need to take in order to meet my math success goal are:

Action Steps	**Target Dates**
a. ______________________	______________________
b. ______________________	______________________
c. ______________________	______________________
d. ______________________	______________________
e. ______________________	______________________
f. ______________________	______________________
g. ______________________	______________________

9. Here's how I will reward myself for meeting my math goal:

SUMMARY

In this section of our road map, I have asked, Do you really have math anxiety? We've seen that much of the fear and discomfort associated with math is often the result of a host of other academic difficulties. You were asked, What does successful accomplishment in math mean to you? I've encouraged you to explore your goals for math success and to develop a plan for achieving them.

CHAPTER

2

THE MATH ANXIETY PROCESS

Mary has been math anxious as far back as she can remember. She hates math. She dreads it. She avoids any contact she could possibly have with numbers. Where did it all begin, I wondered. I asked Mary to write her math autobiography so we could get some insight into the roots of her problem. And there it was. The mystery of her great fears about math lay in front of me in black and white. Mary wrote:

> I remember I was in grade school, and I loved it. But the nuns were very strict. And one day I had to go to the bathroom real bad. I raised my hand to ask, but the teacher didn't wait for me to ask for permission to leave; instead I was called to the blackboard to complete a math problem. And there at the board I lost control of my bladder. It was awful. Right in front of the whole class. I've hated math ever since.

Bob's math autobiography revealed that his math anxiety didn't begin until junior high school. He was taking a prealgebra class and having difficulty understanding several of the concepts. He wrote:

> My teacher became so angry with me, he yelled at me in class, saying I'd never be able to do math. And he was right. Ever since that time math has been my worst subject. I'd avoid it now if I could, but I need it for my major.

John remembered needing help with his math homework in third grade. His older brother began to help him and tried to explain some of the concepts. When John still had difficulty understanding these concepts, his older brother beat him up. John made up his mind at the time never to learn math again. He was 45 years old when he came to me and said, "I'm ready to begin learning math."

Rachel had a lot of difficulty in her fifth grade math class. She wrote in her math autobiography:

> My teacher was so frustrated with my asking so many questions in class, that every time she taught math she had me sit in the hallway outside of class. She said I was a hopeless case and I couldn't learn math anyway.

As unbelievable as these cases sound, they are all true examples of students who have come to me for math anxiety counseling. The techniques I describe in this book helped these students overcome their fears and successfully accomplish their math goals. My strong message to them—and to you—is that *your past negative math experiences needn't continue to burden you*. I encourage you to throw off the shackles of the past. Starting today, tell yourself, "I can and will succeed in math."

REEXAMINING YOUR PAST MATH HISTORY

I've been able to identify several different reasons for the onset of math anxiety in the students with whom I've worked. Let's explore some of these reasons and see how they might have affected you.

EMBARRASSMENTS

Have you every been embarrassed about math? Did you feel shy or mortified when being asked to do math in front of others in class? Many, many students report embarrassing moments related to math, dating as far back as first grade all the way through high school. They quivered with self-consciousness when called upon to answer a math problem or do their times-table in front of the class. One wrong answer, and they felt humiliated. Many feared being called to the blackboard to work out a math problem, standing there all alone, with everyone watching. Public speaking fears and stage fright became associated with embarrassment over math. Other students reported being drilled over and over with flashcards, and how flustered they became when they forgot even the easiest of facts or when they were caught counting on their fingers. Still other students were mercilessly teased by classmates if the answers came too easily and they seemed to enjoy math. They were called names like "nerd," "show off," "bore."

POOR CURRICULUM

I was at a party recently where I spoke with a prominent lawyer in my community. I told her I was doing work with math anxiety reduction. She looked at me and emphatically exclaimed, "Mrs.

ANOTHER CAUSE OF MATH ANXIETY

Thompson, third grade!" She went on to tell me that her third grade teacher, Mrs. Thompson, had introduced "new math" to her. And no matter how hard she tried, she couldn't comprehend this new approach, nor could her parents help. She said it was all over for her in math from that point on. To this day, the lawyer explained, she has trouble with math and still has difficulty figuring out the 15% for leaving tips. She carries with her a calculator with a 15% button for computing tips on her business luncheons with clients. For many students, the introduction of new math was a major stumbling block in their math progress. Other curriculum choices that have adversely affected students include unsatisfactory textbook selection, inadequate prealgebra preparatory courses, gaps in course or unit sequencing, and too fast a pace.

NEGATIVE LIFE EXPERIENCES ASSOCIATED WITH LEARNING MATH

Many students have reported traumatic childhood experiences quite unrelated to math, but somehow the emotional upset of these experiences became associated with math. Phyllis's parents got divorced when she was learning long division, and to this day, Phyllis is distraught whenever she is required to do math, especially division. Mark's family suffered a lot of emotional turmoil throughout his junior high and high school years. His father was an alcoholic, and family life was punctuated by many arguments. His home situation severely affected his ability to concentrate in math during those years, and math still remains an area of extreme discomfort to him. Norm's grandmother died when he was in sixth grade. She had been living with his family and her absence was taken very badly by everyone, especially Norm. He began to fall farther and farther behind in math and did poorly on his math tests. Soon Norm began dreading the thought of ever having to take a math class. He would do anything to avoid math.

When you look back to the days when you first began to have math discomfort or fears, can you identify events in your life that were so emotionally disturbing that they could have become associated with learning math? Or perhaps you experienced an illness or other interruption in your education that caused a critical gap in your math background.

FAMILY PRESSURES AND EXPECTATIONS

Did members of your family ever try to tutor you? Did they ever reprimand or yell at you or show disgust when you didn't seem to comprehend math quickly enough? Did your parents push you to succeed or compare you to a sibling who was a "math whiz"? Or perhaps you had relatives who discouraged your math achievement, telling you they were never good in math either. I've worked with students who had nightmares as a result of being tutored by demanding or overzealous relatives, and I've met others who vowed never to take math because they hated being compared to a sister or brother who had excelled at math.

Margo's father was an engineer who very much wanted Margo to be as competent in math as he was. Every night he tutored her, went over her homework, tested her on her progress. Margo wanted so much to please her father, but when she didn't understand the math fast enough, he often got so frustrated he slammed the book down and walked away. By the time she got to high school, she began to avoid math. Unwilling to repeatedly face her father's disappointment in her, Margo decided to take as little math as possible.

DESIRES TO BE PERFECT

Have you feared math because it seemed so exact to you, that there always appeared to be a right and a wrong answer? And if you got a wrong answer, did you feel that it was a poor reflection on you and your academic abilities? Many students feel like this. They often have a fragile, negative self-image. They don't feel good about themselves, and they try to avoid situations that point out their weaknesses. Math puts them in such a situation. Each time in the past when they were called on in class and gave a wrong answer, it made them feel less worthy, dumb, or unintelligent. They could try and try to work math problems, but when their answers came out wrong, they were left with feelings of incompetence and poor self-esteem. Is it any wonder that they began to avoid math? Has math affected you this way?

POOR TEACHING METHODS

In elementary or junior high school, did you have teachers who really didn't like math or want to teach it or who weren't well trained to teach math? Did you have teachers who couldn't answer your questions or perhaps who put you down when you didn't know the answer to a question? Did you find that your teacher couldn't quite explain the material in a way that you could understand?

One student reported that his fifth grade teacher said that if the class was good all day, they wouldn't have to do their math that day. Math became a punishment for the class if they misbehaved. It turned out that this teacher had been trained as a drama teacher and was quite frightened by math. In the past, many college students who were math anxious chose elementary school teaching to avoid most college math requirements. In doing so, they successfully sidestepped any confrontation with their anxiety, but they still harbored it. Since math is taught in all the elementary grades, these teachers had ample opportunity to pass their math fears and uncertainties on to their impressionable students.

In other situations, teachers have ridiculed students or told them they would never be able to learn math. Teachers rushed through the material or gave few or no problems for homework to reinforce the concepts that were presented. In some cases, students needed to have hands-on math experiences, and the teacher only lectured or wrote on the board but couldn't provide adequate alternative teaching methods to help the students learn.

NEGATIVE MATH GAMES PEOPLE PLAY

Do you ever say things to yourself that seem to block your ability to do math? Your internal mind talk can play an important role in your math performance and could easily be a negative or destructive influence on you. When this talk is detrimental, we refer to it as *negative math games*. These games, if not reversed, can result in a complete loss of self-confidence in math, and self-confidence is one of the most important aspects of achieving math success. The following statements are examples of self-defeatist games students play on themselves. You may say things like: "I was never good in math, so I can't be good now." "Why do I need math anyway?" "Math isn't useful in my life." "I don't work math quickly enough." "Everyone knows how to do it but me." "This is a stupid question, but . . . " "It's too easy, I must have done it wrong." "I got the right answer, but I don't know what I'm doing."

Aside from the negative math games you may play on yourself, there are games that others may play on you. These are statements that others say to you, deliberately or perhaps unintentionally, that negatively affect your ability to do math. Perhaps you sought help on a math problem and the other person responded: "Oh, that's easy"; or "You should know that by now." "You'll never be able to do math." "You'll just have to work harder in math, and you'll get it." "The answer's right in front of you, don't you see it (you dummy)?" Statements like these just serve to make you feel bad and to increase your fears and uncertainties about math. Other games people use include, "You got the right answer, but you did it the wrong way." Since there are many ways to solve a math problem, this game just serves to increase both your self-doubt and your anxiety level. Still others might try to discourage you from learning math by saying, "Why learn math anyway; you'll never need it?"

CULTURAL MYTHS AND STEREOTYPES

Have you accepted the many powerful myths and stereotypes that have surrounded the subject of math? Our society seems to readily accept mathematics illiteracy. Very few people would admit that they couldn't read; but those who cannot do math will find lots of company, and acknowledging their deficiency not only produces no social stigma but generates empathy and commiseration. Furthermore, the popular media tend to portray those skilled in math as intellectually superior and, therefore, strange or different.

Throughout our society parents, teachers, friends, relatives, books, magazines, and television have often perpetuated a system of false beliefs about math. It is believing in these falsehoods, which have no basis in reality, that can stop you from progressing in math. Let's look at some of these beliefs now.

Do you believe that you must have a "math mind" or be a mental giant to succeed in math? Whereas there are people who have extraordinary aptitude and ability in math, as in any subject, you needn't be a genius to become competent. You *are* able to do math. There is no special innate ability that we inherit that relates only to this subject. We are all capable of learning math. It takes time, patience, determination, and a great deal of practice.

Are you convinced that there is a magic key to doing math or that math problems must be worked on intensely until they are solved? Do you think that there is necessarily a best way to solve each math problem? Once again all these are falsehoods. There is no magic key or formula for doing math. There is no one best way to solve any problem. Usually, there are a variety of different ways to find the correct answer. And math problems often are not solved in one sitting. It is important to take breaks and rest during problem solving. Mathematicians may take several days, and sometimes months, to solve a difficult problem.

Have you often thought that you should be able to do math quickly in your head and that only dunces count on their fingers? Any physical model that helps you solve math problems is OK. For example, finger counting may actually show that you have an understanding of what you're doing and that you aren't merely doing math through memorization. As for doing a math problem in your head, this is too much to expect from anyone who hasn't done this sort of thing many times before. Even experienced mathematicians can't necessarily work out new math problems in their heads.

Do you believe that only men are good in math or that careers requiring math are mainly for men? The men that I work with who are math anxious would certainly disagree with you. They feel that men aren't any better in math than women are. Cross-cultural research studies have shown that in some countries female students' achievement on standardized math tests is equal to or superior to males', pointing to the fact that American females' poor representation in fields requiring math is almost certainly a cultural phenomenon (Gray, 1981; Hanna, 1989).

Have you often thought that math is not creative or that math requires only logic and rational thinking but not intuition? Perhaps you've believed that you must always know how you got your answer or that it was always important to get the right answer in math. None of these statements represents reality; all are math myths. Math can be very creative, imaginative, and intuitive. Mathematicians often use their intuition to figure out solutions, and they can't always explain how they arrived at their answers. Problem solving can be a very creative, innovative process. Playing with various methods, testing out what feels right, sleeping on the problem, or brainstorming different solutions can all be fun and inventive. Knowing the precise answers often doesn't matter either since many times answers to difficult problems may only be approximations or "guesstimates."

> *Here, where we reach the sphere of mathematics, we are among processes which seem to some the most inhuman of all human activities and the most remote from poetry. Yet it is here that the artist has the fullest scope of his imagination.*
>
> *Havelock Ellis*
> *The Dance of Life (1923)*

WHAT IS YOUR MATH HISTORY?

So, as you can see, the origins of math anxiety are varied. The following exercise will help you focus on your math history and will help you identify the early roots of your problem.

EXERCISE 2-1: YOUR MATH AUTOBIOGRAPHY

What is your math history? In the spaces that follow, briefly describe your chronological history in terms of the negative and positive experiences you've had with math. Include your earliest memories, as well as memories of how your teachers and your family influenced you in math.

Describe how your family members approached math and describe their attitude toward your math ability. Include a description of how you've dealt with recent situations involving math in other classes, on the job, or in daily life situations. End with a discussion of how math could help you in accomplishing your educational objectives, in earning more money, in choosing a career or in any other aspect of your life. (Use additional paper.)

REASSESSING YOUR PAST

I want to reiterate here what I have stated earlier: ***your past negative math experiences need not continue to burden you***. It is both your responsibility and within your power to take control of your math destiny. In the next exercise, I encourage you to review your math history so that you can understand the roots of your fears. Then I will ask you some important, thought-provoking questions to help you to reevaluate your previous experiences. You can throw off the shackles of the past that have weighed you down so you can make progress.

EXERCISE 2-2: NEW INSIGHTS AND REVELATIONS

When you finish your math autobiography in Exercise 2-1, look it over and see if you can find the roots of your math anxiety. Is it clear where it began? What factors most influenced you? What past attitudes and actions on your part contributed to your current difficulties? Then ask yourself: "Are these conditions still relevant in my life today? Are the messages I carry around about my abilities to do math true today? Can I move further and not let the problems of the past be my perpetual stumbling blocks to learning math? What positive experiences can I build on?" In the space provided, describe the insights you have gained from reading over and analyzing your math history.

PICTURING THE MATH ANXIETY PROCESS

As you can see, unpleasant encounters with math in formative years can be ruinous to subsequent learning. Students who were made to feel bad about math become wary and prejudiced against it, mistrusting their own ability. New experiences in math, seen in light of the old, are tarnished by the troubled past, which only accentuates and reinforces long-entrenched negativity. Bad feelings persist, crippling prospects for learning new material and generating anxiety and self-doubt. Math-anxious people often say things to themselves such as "I'm stupid," "I'll never be able to do math," "I'll fail," and "Why do I need to know math anyway?" Soon a continuous flood of negative talk about math ensues; before long, anxiety, overwhelming fears of failing or looking stupid, and panic set in. Physically, these people may experience nausea, perspire profusely, develop a headache or tight muscles, or exhibit a number of other physical symptoms. Mentally, they become confused or disorganized, make lots of dumb errors, forget formulas they knew, can't think clearly, or blank out entirely. The end result: poor progress, avoidance of math, feelings of failure. I have diagrammed this process in Figure 2-1.

When you deal with math anxiety, you must pay scrupulous attention to what you really think and feel. Negative self-talk seems to be an especially critical habit, which keeps perpetually alive the

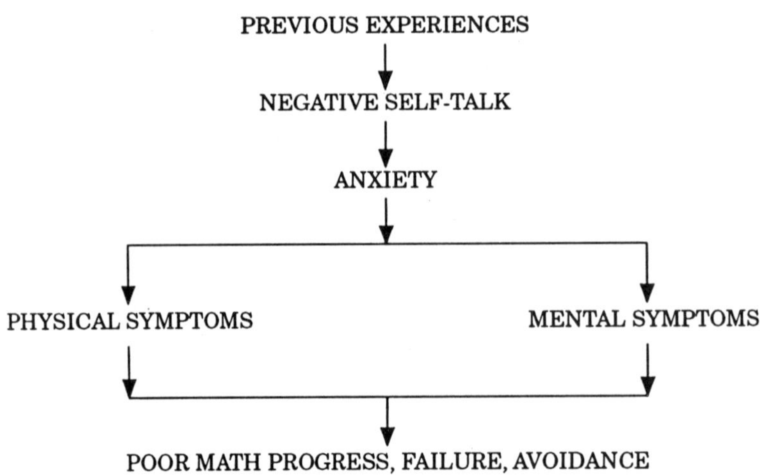

FIGURE 2-1 THE MATH ANXIETY PROCESS

negative experiences of your past and fosters an ever-increasing anxiety level. Let's begin looking your negative dialogue straight on. In the following exercises, you are encouraged to keep a math journal. Writing a journal is an effective means of becoming aware of all your inner mind chatter surrounding math. I encourage you to heighten your awareness of what you are saying to yourself when doing math. What are the situations and ensuing self-statements that trigger your anxiety?

EXERCISE 2-3: MATH JOURNAL WRITING

Starting *today*, begin to keep a math journal about your everyday experiences using math. Use a format similar to the chart entitled "My Math Journal," which appears at the end of this exercise. Some students choose to write their journal in more of an essay format, and this is OK too. But no matter what format you choose, I encourage you to include in your journal the responses to all the questions I have posed here. Using a separate notebook or several additional sheets, continue this journal for at least one or two months. Make at least two entries per week.

Column one of the journal asks, "What situation required me to use math today?" Even if you are not taking a math course currently, there

MY MATH JOURNAL (sample form)			
What situation required me to use math today?	**What I said to myself and how I felt during and after the situation**	**What I've learned about myself**	**What I'm going to do about it**

are many daily life situations in which math is needed. Focus in on these for your journal writing.

Next, respond to the statement in column two of your journal, "What I said to myself and how I felt during and after the situation." Listen carefully to your inner voice and note all your mind talk. Did you say negative statements to yourself about math, about your ability to do math or about your intelligence? Then take a look at your feelings. Did you notice any physical symptoms such as butterflies in your stomach, sweaty palms, or tense muscles in your neck or shoulders? Did you feel dumb or stupid, or confident and assured? Did you blank out, feel panicky, overwhelmed? Did you feel like giving up easily or like staying with the situation, "even if it killed you"? Were you afraid to make a mistake or afraid of being judged? Record all feelings and sensations.

Column three in your math journal is very important. It asks you to review "What I've learned about myself." It is from here that you will begin to find ways to overcome your specific anxieties in math. Here you must objectively look at what you learned from this current math lesson. Did you learn that you are too hard on yourself, that you are saying pretty nasty things to yourself about math and your ability to do math? Does one particular thought tend to repeat itself often and make you feel defeated? Did you notice that you feel unsure of your math ability even when you can do math problems correctly? Did you notice that you block your own progress by putting up an impermeable wall or blockade? Do you give up too readily as soon as the problem seems a little difficult? Do you know how to do the math, but the anxiety makes you uncomfortable? Did you notice that you prefer working alone, in a room with others, in a quiet space, or with music in the background; in the morning, or the evening, or late afternoon? Did you notice that you concentrate better when others aren't watching? What other important things have you discovered about yourself? Write down everything that you learn from observing your current math experiences.

The last column in your math journal asks you to assert "What I'm going to do about it." Having identified what you have learned from this experience, you are now being asked to make a commitment and take some positive actions. Ask yourself: "What relevance does all of this have to my future math success? What can I do to feel better about math? How can I increase my chances of succeeding? How can I structure my dealings with math so I am comfortable with it? What can I say to myself to feel better about my math abilities?" Describe all the steps you are willing to take.

As you continue to keep this journal in the next few months, and continue to progress through this book, more and more ideas will come to you regarding the most relevant steps you need to take to succeed in math. Jot them down; capture all your ideas on paper.

SUMMARY

In this section of our road map, we have investigated the roots of math anxiety and searched into your math history to see where it all began. We have looked at poor teaching methods, embarrassments, poor curriculum, negative life experiences, family pressures and expectations, desires to be perfect, negative math games, and math myths and stereotypes. We also diagrammed the math anxiety process and saw how negative inner dialogue perpetuates it. You were encouraged to begin a math journal of your current math experiences in order to gain important insights into how your thoughts and feelings affect your math success. Throughout the chapter my strong message to you has been: your past negative math experiences need not continue to burden you. So I urge you to throw off the shackles of the past and, starting today, say to yourself, every day, several times each day, "I can and will succeed in math."

CHAPTER

3

LEARNING TO MANAGE ANXIETY

Imagine, for a moment, a virtuoso concert violinist tuning her masterly crafted violin. When the strings are too loose, the violin cannot be played. The violin sounds flat, sour, lifeless. The tension is too weak and, no matter how well trained the violinist, the violin cannot play beautiful, harmonious music. On the other hand, if the strings are strung too tightly, the resulting sound is screeching noise, grating and uncomfortable to the ear and far from pleasing. Just the right amount of tension is needed for the violin to sound its best. And at times during a performance, the violin has to be adjusted and readjusted to maintain this delicate balance. Only through special attention and sensitivity to the fine tuning of the violin can the violinist be assured of continuous success. The same is true in learning to manage anxiety.

We must learn how to finely tune our anxiety level, for anxiety can be our enemy or our friend. With too little anxiety, we feel lazy and unmotivated. We don't push ourselves to perform or achieve. Our memory seems somewhat dull. We don't think as clearly as we could. On the other hand, when we experience too much anxiety, we feel out of control and devastated. We become tense, out of balance, panicky. Our thinking appears confused and disorganized. We feel insecure, inadequate. Our minds often go blank, and we might experience physical symptoms of muscle tightness, diarrhea, nausea, or vomiting. Images of doom and disaster loom over us.

However, as with the finely tuned violin, when we manifest just the right amount of tension, not too little nor too much, we think clearly: our memory is sharp, our perceptions are accurate, our judgments are good. Maintaining the proper amount of anxiety helps us to perform at our best. Thus, the goal of anxiety management is not to alleviate all the anxiety you may experience but to help you learn how to manage and finely tune the anxiety you have.

In this portion of our road map, we will explore the physical aspects of anxiety and how to manage and fine tune it so that you can function at your optimal level.

Let's look at what happens when you experience an anxiety reaction. When you begin to feel anxious, many physical changes occur. Your body reacts as if it were in danger, and prepares for possible fight or flight. For example, the muscles of your arms and legs often tense, anticipating the need for action; the pupils of your eyes dilate, letting in more light to sharpen your eyesight; your heart rate increases to circulate blood more rapidly to the brain and vital organs; and respiration increases to provide more oxygen to the tissues. Physiologically, hormones are produced that activate the sympathetic portion of the autonomic—or involuntary—nervous system. All of these sympathetic reactions would be perfectly suited for survival when dealing with real-life emergencies.

Working out a math problem is obviously not a life-threatening event, but you may be approaching it as if it were. You may begin focusing more and more on fearful, negative self-talk and on the sympathetic bodily sensations. This only serves to increase the anxiety and tension in your body. In a later chapter, we will deal with ways to change the negative self-talk and negative attitudes toward math that feed into this anxiety process and continuously threaten to make it overpowering.

As your anxiety reaction intensifies, the changes in the rate and pattern of breathing tend to be more pronounced. Instead of breathing from the lower lungs, you begin to breathe much more rapidly and shallowly from the upper lungs. You may not even be aware of this more accelerated breathing. As you breathe more quickly, you expel too much carbon dioxide too rapidly. This produces a phenomenon known as hyperventilation, a condition that produces many of the uncomfortable bodily sensations that math-anxious students experience and focus on. Some of the symptoms of this condition include confusion, inability to concentrate, shaking, fatigue, muscle spasms or pain, difficulty swallowing, tightness in the throat, choking sensations, shortness of breath, dizziness, irregular heart rate, numbness or tingling of the extremities, and lightheadedness.

As you become more and more aware of these symptoms, they seem to increase. This process, if not stopped in its tracks, can lead to a total panic reaction in which you may completely avoid learning or doing math, freeze up, or actually get sick when approaching math. The following is the most important principle you need to know for managing anxiety:

> By changing the rate and pattern of your breathing (during anxiety-producing situations) to a deep, slow breathing pattern utilizing your lower lungs, you can bring a sense of calm and ease to your body and mind.

Exercises 3-1 and 3-2 will teach you how to calm yourself physiologically and prevent panic from setting in. Instead of automatically breathing rapidly and shallowly with your upper lungs, you learn to breathe gently, slowly, and fully from your lower lungs. These breathing exercises will produce the following calming effects:

decreased heart rate
decreased oxygen consumption
decreased muscle tension
decreased blood pressure
slowed breathing
increased sense of calm and peace in the mind
increased sense of relaxation in the body

EXERCISE 3-1: THE CALMING BREATH

Steps

1. With your eyes closed, sitting in a comfortable position, slowly and gently inhale, concentrating on filling the lower part of your lungs and expanding your abdomen.
2. Now, very slowly and easily, exhale . . . saying "relax" as you do so.
3. Continue breathing gently and slowly, filling the lower lungs and then very slowly exhaling, saying "relax" and feeling more and more relaxed with every breath you take. Continue for approximately 10 more minutes.

I recommend you practice this exercise each day for as little as 5 minutes and as long as 20 minutes.

EXERCISE 3-2: DEEP ABDOMINAL BREATHING TECHNIQUES

This is an extremely powerful technique during times of panic or anxiety. It should be used for no longer than five minutes at a time. However, if you wish to continue longer, you should switch to the calming breath technique described in Exercise 3-1.

Steps

1. With this method, it is important to breathe from deep within your abdomen. You may wish to put your hand on your abdomen to feel if you are truly breathing at this deep level. Have your eyes closed and sit in a comfortable position. Now, while slowly counting to four, take a long, deep breath, first filling your lower lungs, feeling your abdomen expand, then filling your middle and your upper lungs. Now hold this breath for the count of two.
2. Now, slowly exhale to the count of four while relaxing the muscles of your face, jaws, neck, back, and stomach. And now rest for the count of two.
3. Now begin again. *Inhale* slowly to the count of four. *Hold* to the count of two. *Exhale* to the count of four. *Rest* for the count of two. Continue this exercise for a total of five minutes.

Variation 1: The Extended Exhalation Technique

One variation of the deep abdominal breathing technique is to take a slow, deep breath, feeling your abdomen expand, and hold the air for a moment; then very, very slowly, exhale, counting to ten as you do so. Continue this breathing pattern for approximately five minutes and then go back to the calming breath technique described in Exercise 3-1.

Variation 2: The Gentle Rhythm of Your Breathing

In this technique, you are simply to observe the rhythm of your breathing. Allow your consciousness to become totally aware of your breathing. Notice the temperature of the air, how it is cool when it enters and warm when you exhale. Observe how the air flows into your lungs and then how your abdomen rises and falls with every breath in and out. Continue to focus on this gentle rhythm. Allow it to quiet your thoughts and bring you a sense of calm, peace, and joy. Practice this for about 10 to 15 minutes each day. This is a wonderful technique to relax when you feel yourself tensing up while doing math.

Variation 3: The Extended Sigh Technique

Practice this exercise in a quiet, private place. You may do it standing, sitting, or lying down. Begin by taking a deep, comfortable breath, filling your lungs completely. Then, while exhaling very, very slowly, allow yourself to let out a deep resounding sigh. Let the sigh extend as long as you can. Enjoy the intense feeling of relief this produces. Practice this deep extended sigh 10 to 12 times each time you do it. Use it whenever you are feeling a sense of frustration or high stress.

Be sure to practice at least one of the breathing techniques described in Exercises 3-1 and 3-2 a few times a day, every day, for several weeks. These breathing techniques are particularly beneficial when used before entering a math situation in which you are very anxious. By practicing them daily, you will become more comfortable and familiar with them so that they can be immediately available to you during times of high academic stress.

Our hectic life-style, in general, often keeps our stress level so continuously activated that our minds and bodies seem to stay aroused at extremely high levels for days, maybe weeks. Through practicing relaxation exercises on a regular basis, you will be the master of your anxiety. You will be able to quiet your mind and body and restore homeostasis, your body's internal equilibrium.

Studies have shown that the systematic use of relaxation techniques on a daily basis for 10 to 20 minutes produces many significant benefits related to academic performance:

improved long- and short-term memory recall
increased measured intelligence
improved perceptual awareness
relief from insomnia
decreased anxiety, irritability, and depression
increased emotional stability
improved self-esteem
greater capacity for reaching one's potential
improved academic performance in both high school and college

If you are experiencing a great deal of anxiety over math, you will need to create your own special approach. Do it and get a handle on the earliest signs of anxiety so that your anxiety level can't get the best of you and grow to panic-like proportions. There is no magical formula to prevent anxiety or panic in math. It will take practice, determination, a positive belief in yourself, and a commitment to reach your math success goal.

SUMMARY

Managing anxiety is like tuning the strings on a violin. When the tension is perfectly adjusted, the resulting music is sweet and melodic. When you learn to fine tune your anxiety level to allow just the right amount—not too much, not too little—you, too, will perform like a virtuoso.

In this section of our road map, we probed the physiological changes you experience during an anxiety reaction. We then explored methods for physically calming your mind and body. I encouraged you to practice several different breathing techniques. When you are anxious, you must change from a shallow, rapid breathing pattern to a deep, slow breathing pattern using your lower lungs.

CHAPTER

4

OVERCOMING INTERNAL BARRIERS

One of the psychology instructors at my college once said to me, "I've noticed that how my students deal with math is how they deal with life's problems." As I pondered what he said, I recognized the truth of his comment.

How do you handle problems when they emerge in your life? Exercise 4-1 will help you see how you deal with obstacles as they arise. This fantasy will be of most value to you if you are completely honest with yourself and if you experience it before you read subsequent chapters of this book. Have someone read the exercise to you very slowly and calmly; or perhaps you can read it into a tape recorder and later play it back. This exercise is an adaptation of a brief activity originally described by Susanne Culler in *The Resource Manual for Counselors / Math Instructors,* The Institute for the Study of Anxiety in Learning, Washington, D.C., 1980.

EXERCISE 4-1: THE WALL FANTASY

Arrange to be in a quiet, dark room where you will not be disturbed for approximately 15 minutes. Sit or lie comfortably while you listen to the following fantasy (each set of three dots indicates a few moments' pause):

Begin by getting as comfortable as possible. Settle back, allowing yourself to sink down inside, and gently close your eyes. It is important when you do fantasy work to be in a relaxed and meditative state.

Let's begin by focusing on your breathing for a few moments. Take a deep and comfortable breath, filling your lungs completely . . . hold it a moment . . . and then . . . very, very slowly, let it out . . . slowly . . . feeling a wave of relaxation going from the top of your head all the way down to your toes . . . Good! Now take another deep and comfortable breath, filling your lungs completely . . . hold it a moment . . . and then . . . once again let it out very, very slowly, feeling another wave of relaxation going from the top of your head all the way down to your toes . . . slowly . . . feeling more and more relaxed as you do so.

Continue to breathe slowly, deeply, and regularly. (Pause 30 seconds.) Feel your body becoming more and more relaxed, feeling heavy and more and more relaxed. Allow all the tension to leave your body. (Pause for 20 seconds.)

Now let's begin our fantasy by imagining that you are walking along a road in the country. It is a beautiful day. The sun is shining warmly, but it is not too warm for comfort. The air is crisp and clear and there is a gentle breeze that feels delightful as it glides past your cheek. The dirt beneath you feels warm and soothing to your feet. The fields on both sides of the road are lush, with vibrant green rolling hills. In the distance you can see trees and wildflowers of every variety and color. Oh, it is a beautiful day!

The road you are walking on leads up and down gentle hills. Feel yourself walking. Feel the gentle breeze blowing through your hair. Feel your arms swinging as you walk along.

The road makes a gradual curve and then, suddenly, your path is blocked by an enormous WALL. Walk up to the wall and feel its texture. As you look up, you see the wall stretching so high, it seems to go into the clear blue sky above. To your right, the wall seems to stretch as far as the eye can see. And to your left, the wall stretches so far that it seems infinite. How do you feel? What do you do now? How do you proceed? (Pause for two minutes.)

Now slowly bring yourself back into this room. Bring with you the thoughts and feelings you had during this fantasy.

EXERCISE 4-2: ANALYZING YOUR FANTASY

Having completed Exercise 4-1 these are some questions I'd like you to think about :

What did your wall look like?
What did you do when you got to the wall?
If you managed to get to the other side of the wall, how did you accomplish this feat?
What do you think is the purpose of doing this fantasy?
What is the most important thing you've learned about yourself that can help you to tackle obstacles you face, particularly with regard to math?

Examine how you deal with the wall in Exercise 4-1 and see how it resembles the way you treat problems in your life. This will also give you some sense of how you might be handling math. To overcome life's obstacles we start with an awareness of what we are doing so we can stop our own progress and then make a conscious decision to change in order to prevent these obstacles from standing in our path.

When you got to the wall, what did you do? Did you resolutely turn back? Did you cry, feel stymied, not knowing what to do, and so simply do nothing? Did you look for a hidden way through the wall and find a secret door? Did you find a big tree nearby and use it to climb over the wall? Perhaps you dug your way under the wall or started removing stones from the wall to dig your way through it? Did you blast your way through it with a bulldozer or a large vehicle?

If you turned back and didn't go past the wall, perhaps there were times in life that you have turned back from obstacles, not looking for ways to overcome them. Have you simply and calmly accepted each obstacle that arises in life, content that maybe this is not the way to go? If this is true for you, what does this mean for how you face the challenge of math? Perhaps you tend to give up too easily when math becomes a bit rigorous, confusing, or in any way anxiety-provoking. Or perhaps you tell yourself that math just isn't for you and that you should choose a major that doesn't require math. Or, instead, you may have been generally complacent about the absence of math in your life, not taking any steps to change this situation.

One student I know, Diana, told me that when she came to the wall she was so frustrated she sat down and cried. She said she seemed to treat her math classes the same way. As soon as a math class got a little hard, she'd get depressed and start to feel bad about herself. She'd give up, wouldn't try, and eventually would drop the course. She had a long history of treating math like this. She knew now she needed to do something different to change this destructive pattern.

GETTING THROUGH THE WALL

In working with the "Wall Fantasy" in my classes, I noticed that students who are determined to get to the other side of the wall tend to be more persistent in tackling and overcoming their math problems. Look at what strategy you chose for getting through the wall. Were you slow and methodical; did you blast your way through; did you find a hidden opening; did you jump over it as if you were on springs? Looking at your approach to math, how have you or how can you now apply this method to tackle difficulties in math?

Rob got through his wall by systematically taking out one brick at a time until he made a big enough opening for his body to slip through. He found that this slow, methodical approach worked best for him in math, too. Every day he worked on ten new math problems and five old ones. He went back to basics, reviewing old materials to make sure he had them down pat. Slowly he chipped away at math as he had chipped away at the wall. His confidence began to build and he knew he was establishing a good foundation for himself in math. Soon he was able to tackle easily any new material that was presented in class.

Joyce blasted through her wall. She then realized she often handled problems in life that way—full steam ahead! But she had never thought of tackling math quite like this until now. She decided that math was not going to prevent her from becoming the civil engineer she had always wanted to be. She started reviewing every math book she could get her hands on. She took a minimum class load and focused mainly on studying math. She joined a math study group, got herself a good math tutor, attended a math study skills workshop, and began studying math two to three hours a day. She practiced relaxation

techniques and positive visualization techniques regularly so she could remain calm and positive about math.

Overcoming math anxiety starts with an awareness of what you may be doing now to stop yourself from succeeding. Are you creating your own internal barriers that impede your progress? Are you handling problems as they appear in your path or do you avoid them? Once you realize what you do to handicap your own progress, you can make a conscious decision to take corrective action.

What if you gave up when you came to the wall and you noticed you have a tendency to give up too easily in math? Ask yourself, Do you really want to overcome your problems with math anxiety? Are you willing to deal with your obstacles *now*? Once you are aware of how you stop yourself from succeeding, you can consciously decide to change this situation. I would suggest you begin by reexperiencing the "Wall Fantasy" (Exercise 4-1). Take yourself through this fantasy again, but this time do it differently. Try alternative ways of getting through or past the wall. Experiment until you find the most satisfying way to overcome this major obstacle. Remember this is *your wall,* this is your barrier to your future success. You can do it! You can get through it, over it, under it, or even blast it down, leaving it in rubble.

You may also wish to experiment with other techniques to deal with your internal blocks in math. Journal writing, described in Chapter 2, is an excellent method. Exercise 4-3 is a variation on journal writing, and it can help you to confront word problems, a nemesis for many. You could also talk with a counselor or math teacher who may be able to give you ideas on how to proceed.

EXERCISE 4-3: WORD PROBLEM LOG

Many people fear word problems. Faced with one, internal barriers go up and walls seem to appear from nowhere. As you read the following word problems and begin to work them out, pay attention to how you feel, what you say to yourself, and how this inner dialogue inhibits your progress. Don't worry if you're not sure how to approach these problems—it's not the solutions but how you feel about being asked to solve them that's important. You may be quite surprised at what you learn about yourself.

Sample Word Problems	My Thoughts and Feelings
A business executive left her car in an all-night parking garage for three and a half days. The garage charges either an hourly rate of $1.25 or a daily rate of $24, for whole or partial days. Is she better off paying the hourly or the daily rate?	
Eileen is four years older than Monika, who is two years younger than Philip. Philip's twin brother, Mark, is best friends with Joe who is four years older than he. Joe baby-sits for Tom and his newborn baby sister, Julie. Julie is fourteen years younger than Joe. How old is Eileen?	

SUMMARY

In this section of our road map, we examined how we tackle obstacles along our path. We looked at how we deal with life's problems as an analogy for how we deal with math and what we can learn from this experience.

CHAPTER 5

POSITIVE THINKING IS A PLUS SIGN

DO YOU HAVE AN ATTITUDE PROBLEM?

Not long ago I was a guest speaker in a math class giving a talk on math anxiety. Unexpectedly, a young man named Leonard volunteered a thoughtful and inspiring testimonial:

> I used to do real badly in math throughout high school. I thought math was for the birds. I got so frustrated with it, I couldn't stand it. I began hating it. I felt I just couldn't do it, and I didn't want to even try. I figured I was just a lost cause when it came to doing math. But all that has changed. One day, in a quiet moment, a voice inside me said, "Hey man, you've got an attitude problem." The more I thought about it, the more I realized I was the only one who could change things around. So I made up my mind to look at math like it was do-able and important. I saw an image of myself getting good math grades, getting through school, and having a good life. I saw myself using math to solve everyday problems and feeling good about using numbers. And darned if it didn't work! When my attitude about math changed and got better, I started doing great. I like math now, and my work really shows it.

How do you feel about math? Are you intimidated by it? Do you fear it or hate it? Do you sometimes think it's boring or that you're wasting your time? Do negative thoughts about math keep creeping into your consciousness? What do you say when you're having trouble figuring out a math problem? Do you say things like: "I'll never be able to do math; I'm overwhelmed." "Who needs math anyway? Why bother?" Do you feel that no matter how hard you try, you'll never be able to succeed? Have you ever thought that perhaps your negative attitude toward math diminished your ability to succeed in it?

Educators have known for centuries that, for a student to achieve academic success, it takes more than innate ability, competence, or the desire to learn. The key element in this process is having a positive attitude. A positive attitude becomes the catalyst, the supercharger that propels you along the road toward reaching your math goal effectively.

A positive attitude in math becomes like grade A-1 grease in a well-lubricated engine. It enables the engine to run more smoothly, with fewer breakdowns or jammed gears, and with less vibration. The engine seems to hum along, working at top speed, doing its job and doing it well. Any poorer quality grease would gum up the works, make the engine stick, interfere with its performance and perhaps even totally clog it. Starting today, give yourself A-1 top-rated positive attitudes to lubricate your thinking! The time has come to change your negative attitudes to positive, growth-enhancing, reinforcing ones. Exercises 5-1 through 5-11 will help you do this. Several of them are optional. Complete as many as you need to firmly establish your new positive-thinking approach.

EXERCISE 5-1: POSITIVE AFFIRMATIONS FOR SUCCEEDING IN MATH

An affirmation is a powerful, positive proclamation of something you desire, stated in the present tense. It is a direct, short, simple, active declaration containing no negative words or meanings. It is stated in the "now" as if it is already occurring, even though it might not have happened yet. This is done in order to reflect an outcome that is desired in the present or as soon as possible.

Read, speak, or visualize positive math affirmations to yourself several times each day. This will help you to reprogram and replace negative math self-talk. Some students even record their positive math affirmations in their own voice, with relaxing music in the background, and listen to the recording while driving, running, or going to sleep. This has a very powerful effect!

Some people are bound to say, "This is all well and good, but how do you expect someone to be able to do something he or she really can't do now?" The point is, the skill or capacity to do math, especially if it seems difficult, will come to you much more easily if you begin to believe you're developing it *now.* It's like turning an ocean liner at sea—the ship won't change direction until the helmsman turns the rudder. A positive affirmation is the turning force that sets you off in a new and rewarding direction.

Here is a list of positive affirmations to help you succeed in math. Check the ones you believe would be most helpful on your path to math success. Repeat these statements often to yourself. Remember to substitute these or similar positive affirmations every time negative thoughts about math enter your mind.

_______ **1.** I'm becoming a good math student.

_______ **2.** I'm learning more math each day.

_______ **3.** I'm capable of learning math.

_______ **4.** I have good abilities in math.

_______ **5.** I allow myself to relax while I study math.

_______ **6.** I remember more math each day.

_______ **7.** I am relaxed, calm, alert, and confident in math.

_______ **8.** My math improves every day.

_______ **9.** I can understand math if I give myself a chance.

_______**10.** I enjoy math more each day.

_______**11.** I like math because it's useful in everyday life.

_______**12.** Working out math problems is fun.

_____**13.** Math is more and more exciting each day.

_____**14.** Math is creative.

_____**15.** Math is stimulating.

_____**16.** My way of doing math is a good one.

_____**17.** Math helps me to get to where I want to go.

_____**18.** Math is my friend.

_____**19.** I'm feeling better about math.

_____**20.** Math methods help me solve everyday problems.

_____**21.** (Add your own positive affirmations.)

_____**22.**

_____**23.**

EXERCISE 5-2: POSTER YOUR WALL

Surround yourself with positive thoughts and affirmations. Choose your favorite affirmations from the list in Exercise 5-1 or create your own. Then, with brightly colored markers and letter-size colored paper, put each of these positive math statements on a separate sheet. Write them out in large letters and create a fancy border around each poster. Make them as pretty as you can. Have fun with them. Many of my students get a big kick out of this activity. Some of our community college students put flowers on them, or smiling faces or fancy designs. Make at least three or four of these posters. Hang them in the place you usually study. Paste them on your refrigerator, on your telephone, on your mirror, in your car, or anywhere that you can see them during the day. Make them visible—they are reminders of your growing positive attitude toward math. My students believe it would have been wonderful if these posters had been prominently displayed, long ago, on the walls in their elementary and junior high school classes.

EXERCISE 5-3: CHALLENGING NEGATIVE SELF-TALK (*optional*)

Every time negative statements about your abilities or about math come to mind, take note of them and then immediately change or challenge them. Challenge whether your negative math self-statements are reasonable. Ask yourself these questions:

1. Is this self-talk actually hurting or helping me with math?
2. Does this self-talk affect how I feel about math and my ability to do math?
3. Does it affect my desire or motivation to do math?
4. Is my self-talk based on fact or fabrication?
5. Is this statement distorting reality? If so, what is the objective reality?
6. Is this statement 100% true? Is there room for doubt?
7. What evidence do I have to prove that this statement is true?
8. Even if it were partially true, does that make me, or math, bad?
9. What would be a more reasonable statement that I could say to feel better about math?

CREATING PERSONAL METAPHORS

In their musical "South Pacific," Rogers & Hammerstein wrote a delightful song whose lyrics, on the surface, make no sense: "I'm going to wash that man right out of my hair, and send him on his way." Obviously, the composers did not intend these lyrics to be interpreted literally, yet we understand their meaning precisely. Moreover, the image created by the lyrics conjures up a complex set of emotions that most of us have experienced—the notion of a cleansing ritual to get rid of bad feelings and make a positive change.

The composers used the classic literary device known as a metaphor. A metaphor is traditionally defined as a word or phrase that denotes one thing but implies or suggests another. Metaphors are uniquely valuable to us because they help us to develop new insights and ways of looking at our human experience. Through their powerful imagery, metaphors inspire us and lead us on to greater and greater heights and exhilarating emotions. Rogers and Hammerstein again, in their production "The Sound of Music," used an uplifting metaphor in the lyrics, "Climb every mountain, ford every stream, follow every rainbow, till you find your dream." It wouldn't be humanly possible to follow these directions verbatim, yet the implied message—to accept challenges, to seek new experiences, to never give up, to live life fully, all in pursuit of our goals—appeals to us at a deeply personal level.

To the extent that we act on this message, the metaphor has effectively taught us and changed our ideas. Metaphors help us organize our thoughts and give us a direction for our actions. Not just composers and poets, but philosophers and prophets alike have intuitively known and used the power of the metaphor. From Plato's teachings to those of Jesus and Martin Luther King, the metaphor has been a tool for changing our attitudes and influencing our behavior. And it still works!

Exercise 5-4 will help you create your own metaphors to change your negative attitudes and think more positively about math and your ability to do it. In this exercise I ask you to develop a series of positive thoughts, images, and metaphors for guiding your math success.

First, I want you to list positive adjectives or nouns that come to mind when you think of succeeding in math and overcoming your math fears and anxieties. Some examples include: conquer, win, succeed, enjoy, innovative, creative, informative, triumphant, victory, important, fun, worthwhile, useful, helpful.

Second, make a list of inspirational images that accentuate these ideas. Examples of such imagery include: guiding light, beacon, buoy, lighthouse, director of a play, completing a puzzle, reaching the finish line, survival, champion.

The third section of this exercise asks you to describe the growth-enhancing emotions you would like to experience. Students who are overcoming math anxiety have listed feelings of confidence, relaxation, assuredness, excitement, happiness, thoughtfulness, calmness, courageousness, inner strength.

You are now ready to write your own positive metaphors. Section four asks, What metaphors would inspire you and direct your actions on the road to success? What positive ideas, images, and emotions can you put together to create your own special metaphors? Here are some metaphors that have helped other students on their journey:

Reaching for the stars
Knowing the ropes
Seeing the light at the end of the tunnel
Weathering the storm
Gliding along in cruise control
Going full steam ahead
Bulldozing my way through
Sailing through with flying colors
Water rolling off a duck's back
An award winning performance
Going for the gold
Hitting a home run

EXERCISE 5-4: CREATING PERSONAL, POSITIVE METAPHORS

1. List at least ten positive adjectives or nouns that you can associate with success in math.

2. List at least seven inspirational images that come to mind that you can relate to your conquest of math fears and success in math.

3. Describe here the growth-enhancing emotions you would like to experience as a result of overcoming math anxiety.

4. Now, using the positive ideas, imagery, and emotions identified above, develop your own special positive metaphors.

5. Once you create your own personal metaphors, use them to inspire and guide you on your path to math success. Picture the images in your mind, carrying around with you, everywhere you go, the positive feelings they produce.

6. Write the metaphors on separate 3 × 5" cards and post them in the place you study.

PUTTING ENTHUSIASM INTO YOUR LIFE

Enthusiasm can make anything we do 1000 times better. It livens up our lives and our activities, so that we start to feel better about ourselves and what we are doing. It gives us the "oomph," the energy and the forward momentum to achieve anything we set our sights on. When students become enthusiastic about a subject, their motivation increases. They start to see the subject as their ally and not as their enemy. Things they never noticed before become more meaningful to them, and their learning increases.

Here are six steps to help you become more enthusiastic about math. Why not give them a try?

1. DELVE INTO MATH MORE DEEPLY

To get more enthusiastic about anything, you need to increase your exposure to it; explore it, view it from different angles. Start asking more questions, let your curiosity fly. You might search out different math books that further explain the topic you are learning about and that give you a better and deeper understanding of it. In addition to increasing your enthusiasm, it is a great study-skills strategy.

Begin to pay attention to the usefulness of math in your everyday activities. Look for math games and math puzzles. Seek out interesting math facts or curiosities. Look for patterns and relationships; use your imagination. See how math affects every aspect of our lives. Math is amazingly pervasive.

The word *statistics* may have an ominous ring to it, but an astonishing number of average Americans immerse themselves with relish in the batting averages, rebound percentages, and pass-completion ratios of their favorite sports heroes. Probability theory may seem hopelessly esoteric to you; the casino owners in Las Vegas are counting on that reaction on your part as you part with your money. You may be consumed with the task of equitably dividing up a pizza at a party, yet be blissfully unaware that the same theoretical methods apply to apportioning congressional districts.

Like the infinite hexagonal variety of snowflakes, mathematical patterns in nature surround us with beauty and mystery. The spirals on the surface of a pine cone or pineapple, the arrangement of leaves around a plant stem, the growth spirals of seeds in a sunflower or sections of a chambered nautilus shell all follow the same pattern of elegant mathematical symmetry (known as the Fibonacci sequence). Artists in all cultures have duplicated these patterns in their work for thousands of years.

Fractal geometry describes the shapes of natural objects and boundaries, and modern computer graphics have translated this complex theory into hauntingly beautiful images of imaginary mountains, coastlines, and fantastic shapes.

We are everywhere surrounded and awed by the power of mathematical consistency in an ordered universe. From predicting election results to predicting earthquakes; from designing interconnected telephone systems to analyzing the interconnectedness of the human nervous system; math is, at the core, the one indispensable tool.

Take time to answer these questions for yourself: Why is math so important? What do you believe are the five most essential uses of math in our society? Can you describe at least five ways math is useful in your everyday life? What are three possible ways math can help you in your career choice? Chapter 10 will help you delve into the importance of math in your life even further.

2. EVERY TIME YOU DO MATH, LIVEN IT UP

Enthusiasm, or its conspicuous absence, affects everything you do. Adding it to your work with math will make you feel better. And, even if at first you don't feel like being enthusiastic—*fake it*! It's strange how the mind works. If you start to act enthusiastic, motivated, full of positive drive to do your math, guess what? Your mind starts to believe it and, before you know it, you are feeling better. So go enthusiastically to math class. Really want to be in class and to be learning the material. Listen and become intrigued, take lots of notes, ask questions with interest and motivation. Do your homework with gusto. Put vitality into your studying. *Don't* grumble to yourself, "I have to take math." Instead, tell yourself "I choose to study math—it's a great opportunity." And say it with life! Remember, it's OK to fake it at first because, pretty soon, it will all be true for you!

3. LIVEN UP YOUR ATTITUDES ABOUT YOURSELF AND MATH

All great successful people know that you are what you think you are. Albert Ellis, in his book *Guide to Rational Living,* tells us that it is our thoughts that affect our feelings and they, in turn, affect our actions. So never underestimate the power of thought.

It is your thought power that directs you in your pursuit of life. If you think you're stupid, if you think you can't do math, if you think you'll never succeed in math—*you're right.* You are what you think you are.

So, beginning today, think positive, enthusiastic thoughts about yourself and about math. Liven up your attitudes. Make them positive and encouraging. Let an optimistic, radiant glow gradually build inside you, a feeling based on the thought, "I can do it!"

4. GIVE YOURSELF PEP TALKS

You are what you think you are—thinking makes it so! So always think positively about yourself and your ability to do math. Never put yourself down or sell yourself short.

Your thoughts, whether positive or negative, grow stronger and more powerful when fueled with constant repetition. You must vaccinate yourself against thinking failure!

What is most important is not how intelligent you are, or what your IQ is, or how much brain power you have. What is really important is how you use what you have. It is not brain power but *thinking power* that counts on our road to math success. The thoughts that guide your intellect are more important than how bright you are. So tell yourself every day, "My positive thoughts and attitudes are more important than how bright I am."

Carry a winning attitude around with you in school, in math class, when doing homework, when studying math. Practice positive pep talks at home, in the car, in bed, in front of the mirror. Discover all the reasons you can for figuring out math problems and why it's fun to solve them. Explore how you can be a winner in math. Use your brain power to help search out and create new and better ways to understand and succeed in math. Tell yourself at every opportunity: "I can do math. I'm a great student. I'm a winner!"

Design your own "Math Winner's Ad Campaign" to use in your daily pep talks. This is a technique for selling math and yourself to yourself! It is like the copy (the words) in any advertising campaign or commercial used to sell a new product. Well, math and you are the promising new product. I'd like you to buy into this wonderful product and discover the joys of being a winner in math!

Carolyn, a sophomore who is overcoming math fears, wrote this pep talk:

> "Now available—a breakthrough in math students—new and improved! ME!!
>
> "You'll find more to like than ever before. Candidly, I'm bright, uncommonly talented, fabulously motivated, and uncompromisingly determined—everything you've come to expect in a proven math success story. But hold on—there's even more. As a further convenience, I can learn math quickly and readily. Simply stated, I know my basic math facts. But that's just part of the story—I'm starting to really understand

algebraic equations. Sounds incredible? And that's not all. I know my formulas and I can easily work out homework problems. As if this weren't enough, my memory is awesome, especially when I stay calm and relaxed. But I don't stop there. You'll be glad to know I'm persistent in my desire to learn and comprehend math.

"The result? I AM A MATH WINNER! Absolutely! Certainly! Fantastic, I admit, but true. Seeing is believing! What's more, satisfaction is assured. And I've saved the best news for last—I come with a lifetime guarantee to succeed in math."

Exercise 5-5 shows you how to develop your own special math winner's "promo." Once you have fashioned it, say it to yourself every day, two or three times a day. It has a ironclad guarantee of success!

EXERCISE 5-5: THE MATH WINNER'S AD CAMPAIGN

1. Think of something positive you can state about yourself in each of the following areas. Don't be modest. Identify what you feel good about, what you do well in, what you feel proud of, what might "endorse" you to others.

My determination to succeed in math ______________________

My motivation to do well ______________________

My intelligence level ______________________

My ability to understand and comprehend ______________________

My memory ______________________

The math skills I already have ______________________

My "stick-to-it-tiveness" ______________________

My desire to learn ____________________

My ____________________

My ____________________

My ____________________

2. Now write the ad copy to be used in your campaign for selling yourself to you! Talk directly to yourself. Include the positive selling points you identified in Part 1 of this exercise. Be direct. Focus only on the positives.

3. Make at least two copies of this ad campaign. Post one in a place where you can see it often. Carry the other around with you so you can read it frequently.
4. Practice saying it aloud, each day, in front of a mirror.
5. Read your ad campaign silently several times every day and before going to bed.
6. Read it before working on math or prior to attending math class. It will boost your spirits and your confidence.

---

5. USE THE LANGUAGE OF A MATH WINNER

Whenever you talk about math and your ability to do math, use positive, encouraging terms. Be optimistic in everything you say and do. If someone asks you how you like math and you say things like "Math is my worst subject," or "It's a constant struggle," or "Don't ask," you are only making yourself feel even worse. So, instead of responding negatively, say things like "I'm starting to really like math," or "Math and I are becoming friends," or "I feel great about it," or "My math ability is getting better each day."

Every chance you have, say good things about math. Never put yourself or math down. When you begin to use the language of a math winner, you will be surprised at how rapidly you'll start to feel better. Tell yourself and the world how great you are feeling about math, what a worthwhile subject it is, how useful you are finding math, how it's helping you get to your career goal, and what fun it can be. By using the confident, assured language of a winner, pretty soon you will begin to feel enthusiastic about math and about your potential to succeed in math.

EXERCISE 5-6: DEVELOPING YOUR PERSONAL WINNING LANGUAGE (*optional*)

List five positive statements you could say about math now if someone asks you, "How do you like math?" Be sure to use words, phrases, or metaphors that signal hope, promise, joy, cheerfulness, satisfaction, lightheartedness, accomplishment, and victory.

a. ____

b. ____

c. ____

d. ____

e. ____

Use these statements whenever talking about math to others, and say them with enthusiasm!

6. DEALING WITH "MATH DOWNERS"— DON'T LET NEGATIVE THINKERS PULL YOU DOWN

Don, a sophomore, was surrounded by friends who discouraged him from taking any more than the minimum required level of math. When he decided to take an advanced level math course, they began to tease him mercilessly: "You're a glutton for punishment"; "You're going to become one of those nerds, like the ones we avoided in high school":

"Here comes Mr. Brains"; and "Why struggle, when there are so many more fun courses around?" Don felt torn between studying math and keeping up his relationship with his "math downer" friends. The comments from his friends eventually became toxic to his enthusiasm. He started feeling bad. He questioned why he needed math anyway. He skipped doing homework to spend more time with his friends and, pretty soon, he stopped coming to class. Don began to feel like a failure and blamed it on math.

Math downers are everywhere, and they seem to have a knack for trying to sabotage the advancement of those who like math. You must protect yourself from people who put you and math down and could potentially feed into your own uncertainties and insecurities about doing math. The views of these people can be deadly to your progress. They can destroy your plan for successfully accomplishing your math goal. They can throw you off track and prevent your journey from reaching a successful conclusion.

So be very cautious. Don't associate with people who *speak* negatively about math. If you must be with these people, don't talk with them about math. Change the subject. Tell them you'd rather not discuss it. And make it a rule: Never accept advice from people who are down on math. Take anything they say with a grain of salt. Use their negative opinions only as a challenge to prove they are wrong about math. Don't let them pull you down. Don't let them dampen your enthusiasm.

EXERCISE 5-7: CAUTION: BEWARE OF NEGATIVE MATH THINKERS (*optional*)

Who are the people who bring you down or discourage you in math? Identify not only those people who say negative things about math, but those significant people in your life who just don't seem to be supportive or interested in your math success. It may not only be the negative words they use or the lack of supportive words, but it may be their body language or tone of voice that communicates that they are math downers. In the space below, identify those people you should be cautious of when it comes to speaking about math. You may use only their initials if you feel more comfortable.

---

7. BUILD A SUPPORT GROUP OF POSITIVE MATH THINKERS

Whenever possible, surround yourself with positive math thinkers, people who like math and are succeeding in it. Attach yourself to these people, let them become part of your math support group. Become their study partners, ask them for advice on homework or points you missed in class. I encourage you to broaden your math support group. A support group is defined as a group of people who support you in the direction you want to go. This will make a world of difference in increasing your enthusiasm for studying math.

Moreover, don't forget that you may have an incredible resource already at your fingertips. If you are now in an instructional program, the teachers, counselors, advisors, and math department personnel at your institution are there to assist you, to help you solve your problems and further your education. Don't hesitate to use their services. Even if there is, as yet, no formal course or workshop for dealing with math anxiety, your mentors may have a wealth of experience with it and will certainly do everything within their power to further your progress. Make them an active and involved part of your math support group.

EXERCISE 5-8: IDENTIFYING YOUR MATH SUPPORT GROUP

In the space below, identify all those people who are positive math thinkers in your life. List those whom you are certain enjoy or love math and think it is fun and exciting. List those people who think it's great that you are working on overcoming your math fears and anxieties. Write down the names of people who are encouraging your progress on the road to success in math. Even if at first you can't think of many people, look around, reach out to others, search for at least ten people who could be part of your positive math thinkers support group.

1. ____________________ 6. ____________________

2. ____________________ 7. ____________________

3. ____________________ 8. ____________________

4. ____________________ 9. ____________________

5. ____________________ 10. ____________________

BELIEVE YOU WILL SUCCEED

Whether you believe you can do something or believe you can't—you're right!

Henry Ford

All of our experiences in life are viewed through our own belief-filtering system. If you believe in something, whether it is good or bad, it becomes true for you and, in effect, it becomes a reality. It is like a self-fulfilling prophecy. Once we believe we can do something, we start behaving in ways to make it happen. Our beliefs are like magnets. Once you believe that something is possible, your mind seeks ways and means to assure it will definitely happen.

The belief in success is the one great driving force behind all successful students. Believe you will succeed in reaching your math goal and you will! Having a positive belief creates the energy, the momentum, and the means needed to accomplish your goal. It fills you with vitality and vigor to charge ahead and to creatively deal with any obstacle that enters your path.

Beliefs can be very empowering in our lives. They tap into the richest resources deep within us. It is belief that activates your mind to find constructive ways and alternatives to reach your goals. Develop a positive belief system about math and success is sure to follow. Negative beliefs stop you in your path.

EXERCISE 5-9: REWRITING DISEMPOWERING MATH BELIEFS

In the left-hand column below, I've listed some common disempowering beliefs that often result in negative feelings and attitudes about math. Add your own disempowering beliefs to this list. In the right-hand column, counter each of the beliefs with a reasonable positive one. I've provided some examples. Use the questions given in Exercise 5-3 to check for reasonableness.

	Disempowering Math Belief	Reasonable Math Belief
1.	Math should come easily to me.	Mathematicians work hard at doing math so why should it come easily to me?
2.	I should do math perfectly.	Even Albert Einstein made computation errors in math.
3.	I should be thoroughly competent in math.	Competency comes through patience, steadfastness, and diligence. They all take time.
4.	There's a right way to do math.	There are lots of OK ways to do anything, including math.
5.	I'm dumb when it comes to math.	I'm bright and resourceful and can learn anything I choose to.
6.	Math is only for scientists.	Everybody uses math—*everybody*!
7.	No one in my family ever succeeded in math, so why should I?	I'm an intelligent and capable person; I've succeeded in a lot of things in my life, why not math?
8.	Algebra is not useful in my life.	
9.	Math is only for geniuses.	
10.	(Add your own disempowering beliefs)	
11.		
12.		

EXERCISE 5-10: BELIEVE IN YOURSELF

Many students can easily describe why they believe they haven't been able to succeed in math. In this exercise, I would like you to change this process. I would like you to identify and list the reasons why you believe you *can* and *will* succeed in reaching your goal.

I believe I have the ability to learn.
I believe I am a hard-working, motivated student.
I believe I can learn algebra (geometry, calculus, etc.)

____

____

____

____

____

____

____

EXPRESSING LEGITIMATE MATH RIGHTS

Acting on legitimate personal rights is an essential part of learning to be a more assertive person. Similarly, expressing your legitimate math rights is important in affirming both your desire and your ability to achieve math success. Sandra Davis at the University of Minnesota adapted an assertive bill of rights approach for the math anxious.

The list of rights in Exercise 5-11 goes further and constitutes a set of guidelines that may be helpful to you in most situations dealing with math. This is only a partial list, and you may want to add to it. I encourage you to act upon these rights, but not to follow them blindly. I suggest that you evaluate these rights, change them in ways to suit you, and use them in conjunction with good personal judgment.

EXERCISE 5-11: IDENTIFYING YOUR MATH RIGHTS (*optional*)

Read aloud the following math rights. Do you agree with them? What other rights should be included in this list? Check off the rights you are willing to act upon. Make a commitment to yourself, today, to begin assertively acting upon your legitimate math rights!

_____ I have the right to enjoy math.

_____ I have the right to achieve my math goal.

_____ I have the right to ask questions of my math teachers.

_____ I have the right to ask "Why?"

_____ I have the right to say "I don't know" or "I don't understand."

_____ I have the right to seek help in learning math.

_____ I have the right to be listened to and taken seriously when I ask for math help.

_____ I have the right to learn math at my own speed.

_____ I have the right to see myself as a capable individual.

_____ I have the right to make mistakes in math and to learn from those mistakes.

_____ I have the right to protest unfair treatment or criticism when I'm doing math.

_____ I have the right to be treated as a capable human being by those who teach me math.

_____ I have the right to assess my math teacher's ability to teach me.

_____ I have the right to seek out the best math instruction possible.

_____ I have the right to positive self-regard and a positive self-image, irrespective of how I do in math.

_____ I have the right to remain calm and confident when doing math.

_____ I have the right to work toward achieving success in math.

Other math rights include:

____

____

____

SUMMARY

Maintain a positive attitude toward math and your ability to do math—this is the supercharger that will propel you along the road to achieve your math success goal. Utilize positive math affirmations, create personal guiding metaphors, write the math winner's ad campaign, use the language of a math winner, build a math support group, and change disempowering math beliefs to more reasonable ones. These are just a few of the powerful strategies offered in this section, strategies that encourage the development of positive attitudes, positive thinking, and increased enthusiasm toward doing math. I encourage you to tell yourself at every opportunity that you can and will succeed in reaching your goal. Tell yourself that you are a winner and that you can do math. Every day say, "My positive thoughts and attitudes are most important."

CHAPTER

6

WINNING WITH "SUCCESS IN MATH" VISUALIZATIONS

I believe I have discovered the one great, moving, compelling force which makes every man what he becomes in the end.

This, I believe, is the greatest force in the universe. I believe all other causes are secondary to it. It is so powerful that the slightest human effort cannot be put forth until it has done its work; and if it should suddenly be annihilated from the world, all activity would come to a standstill, and humanity would become a mass of automatons moving about in meaningless circles.

This force is not love; it not religion; it is not virtue; it is not ambition—for none of these could exist an hour without it. . . . it is imagination.

Clarence Budington Kelland
(1881–1964)

Your imagination is the creative expression of your mind. It helps produce your present and future reality. Once you understand this marvelous visionary ability within your mind, you can use your visualization skills to construct an environment conducive to achieving math success.

We all visualize or see images in our minds all the time. Take worry, for example: Do you know anyone who doesn't? Of course not. Everyone worries at one time or another. But what do we do when we worry? We imagine the worst thing that could go wrong or that could happen in a particular situation. We put energy into picturing the negative.

Bob worried that he wouldn't find a math teacher that he liked. He was always afraid that his teacher would go too fast, wouldn't explain things sufficiently, wouldn't understand him, and on and on. His constant worrying exhausted him emotionally and interfered with his ability to enjoy any math class he attended. Bob, with his negative mental images, succeeded in creating his own negative reality. Because of his strong visualization powers, he always saw his teachers in the worst possible light.

In this chapter, we will work with your visualization powers. But rather than allow you to put your energy into the negative and visualize the worst things that can happen, I will ask you to visualize the positive and to harness your creative energies to realize your fullest potential.

Much research illustrates the benefits of visualizing or mentally practicing an upcoming event. One such example comes from the sports world. A well-known study (Richardson, 1969) shows how visualization can improve the free-throw scores of basketball players. The research was published in the *Research Quarterly* by the Australian psychologist, Alan Richardson. It demonstrates the effects of "symbolic rehearsal" of an athletic activity without any large muscle movements.

Richardson randomly chose three groups of students, all of whom had never practiced visualization techniques. One group of students was asked to practice free throws every day for 20 days. A second group was asked to make free throws on the first day and again on the twentieth, without any practice throwing basketballs in between. The last group was also asked to shoot free throws on the first and the twentieth days. But, in addition, this group was instructed to do something quite different. Participants were asked to take 20 minutes each day and visualize themselves sinking baskets. As in real life, when these students saw themselves miss a basket, they were encouraged to work to improve their aim on the next throw.

The results of this study are fascinating. Richardson found that the first group, the students who practiced free throws every day for 20 days, improved their ability by 24%. The second group, who threw baskets the first and the twentieth days only, as you might expect, showed no improvement at all. The last group, who imagined sinking baskets, improved a whopping 23%! Thus, although they lacked actual physical practice, the last group improved almost as much as the group that had practiced for 20 days. Amazing results—and group three might have done even better, if not for one student. This student could visualize the basketball court, but each time he imagined bouncing the ball, the ball stuck to the floor of the court. From this experiment, Richardson concluded that it is important to control or precisely structure our imagery. It is best to imagine smooth sailing and the overcoming of any obstacles when using visualization techniques for positive goal achievement.

PROGRAM YOURSELF TO SUCCEED

Emil Coue, a nineteenth century French pharmacist, once wrote that "the power of the imagination is greater than that of the will." He explained that you may want something very badly, or will it to happen, but it won't happen if you can't imagine it. You must see it happen in your mind's eye. You must really imagine it being true for you.

Programmed Positive Visualization (PPV) is a technique to help you visualize very clearly whatever changes you want to make in your life and to see your life as you want it to be. It is the deliberate use of the power of your imagination to create your own positive, desired reality. This method can help you consciously change or reprogram a thinking pattern that might have hindered you or, at best, did not help you achieve all that you desire.

PPV is an incredibly powerful tool when used to improve your math abilities. You might rehearse an upcoming math exam, imagine yourself calm and clear in math class, mentally practice asking your teacher for help outside of class, or perhaps see yourself solve equations and difficult problems with ease. You *can* program yourself to succeed in math!

Here are guidelines to set up your own programmed visualization:

1. Decide what it is you'd like to do or what situation you'd like to improve. State clearly to yourself the math goal you would like to accomplish.
2. You must truly want what you imagine. The fewer doubts you have, the greater your visualization power.
3. Some of the relaxation techniques you practiced in Chapter 3 are particularly useful here. When you visualize, allow yourself to be open, positive, and deeply relaxed.
4. You are to picture yourself doing exactly what you want to do, achieving precisely what you want to achieve, creating just what you want to create.
5. You are to build a sequence of positive, powerful steps or events in your mind that portray the fulfillment of your goal. Have each step clearly take you in the direction you want to go. Don't imagine difficulties or failures. Picture yourself overcoming all obstacles. Visualize only the progressive movement toward success and achievement.
6. Your visualization may be even more effective if you are aware of the sights, sounds, tastes, smells, and the feel of what you visualize. The more complete and lifelike the experience, the better. Some people find that they don't actually see images when they visualize, but they can feel, sense, or have the impression or sensation of the images.
7. Choose affirmations that help you feel good and reinforce your progressive movement toward achieving math success. You may choose some positive statements from Chapter 5 or make up new ones that seem more appropriate. State your affirmations to yourself over and over again during your visualization.
8. Trust and believe that what you want will be yours. Push any negative thoughts away. Believe that you can and will attain the success you work toward and visualize in your life. Do not discuss your visualization plan with anyone. Avoid exposure to other's doubts or negative thoughts that could hinder your success.
9. Once you have designed your visualization, decide on a specific time each day when you will take a few minutes to relax deeply and

go through this imagery in your mind. I usually recommend that first you take yourself into a comfortably relaxed state and, once there, begin to repeat your positive affirmations to yourself several times. Next, I suggest that you begin to visualize a gradual progression through the sequence of steps leading to your goal, and end with the image of your goal successfully completed.

10. Each day, no matter where you are—in your car, on your way to class, exercising—visualize your goal and create positive images of achievement in your mind. Always picture that goal as exciting and stimulating. See yourself as successful *now*!

11. Continue this process of relaxation and visualization each day until your math success goal becomes a reality for you. You may then want to repeat this process with a new goal and visualization.

EXERCISE 6-1: "SUCCESS IN MATH" VISUALIZATION

Find a quiet place where you won't be interrupted. Sit comfortably with your eyes closed. Focus on the gentle rhythm of your breathing. Breathe slowly, deeply, and regularly. With every inhalation, feel your abdomen rise, and with every exhalation, feel it fall. Experience all the tightness and tension leaving your body with every breath you exhale. Relax more and more. Go deeper and deeper into a calm and serene state.

Now, with every breath you inhale, say to yourself . . . "I am" . . . and with your exhalation, say "relaxed." As you breathe in, say . . . "I am" . . . and as you breathe out say "relaxed."

Continue alone for five minutes.

Visualize these positive math affirmations as true and say to yourself (pause five to eight seconds between statements):

Deep within my mind, I can see and experience myself relax while doing math.

Deep within me, I can see and feel myself comfortable and confident while in math class.

Every day in every way, my ability to do math is improving.

I enjoy math more and more each day.

I allow myself to remain calm, confident, and comfortable while working out math problems.
Each day, math is easier for me.
I can clearly visualize succeeding in math.
Every day in every way, math is more fun and exciting.
Deep within me, I know I am truly capable and competent to do math.
Deep within my mind, I see and experience the attainment of my math success goal.

Now I would like you to imagine that you are about to do math and that you have your books and notes before you. You have plenty of paper and pencils, and you know you have everything you need. You take a few deep comfortable breaths and calm yourself before beginning to study and work out math problems. See yourself calm, relaxed, confident. Say to yourself, "I remain calm and relaxed while doing math."

Then visualize yourself concentrating completely on math, your studying progressing well. See yourself understanding your math and having the concepts coming easily and readily to you. Tell yourself, "I can understand math if I give myself a chance."

Picture yourself reading over your math notes. Envision yourself reviewing all the important points and quizzing yourself on the concepts. Tell yourself, "I review all the important concepts over and over again to fix them in my mind."

Imagine concentrating on your work, alert, interested, and enthusiastic. Picture yourself working out math problems, the easy as well as the very difficult and challenging ones. Tell yourself, "Working out math problems is fun."

Imagine yourself, after working for about half an hour, taking a five-minute study break to refresh your mind and energize your body. Stretch your whole body, reach for the sky, bend down and touch your toes; move around, get some water to drink or an apple to eat. After your break, settle back into your chair, close your eyes for a moment and take some deep, relaxing breaths. Then visualize yourself resuming study with renewed vigor and confidence. Say to yourself, "Math studying is going well for me."

Now let us conclude your "success in math" visualization by imagining the long-term benefits of confidence and competence in math. Imagine achieving your math success goal. Take the next few minutes to visualize mastery of your goal. Delight in your math success *now*!

How do you feel about yourself and your abilities? How would life be different if you succeeded in reaching your goal? What other benefits would follow as a result of this? Focus on those benefits now! Say to yourself, "Every day in every way I am working toward achieving my math goal."

Conclude this visualization exercise by saying, "Deep within my mind I visualize and experience myself achieving my math success goal."

EXERCISE 6-2: REHEARSING A FORTHCOMING MATH SITUATION

Do you get nervous before math class? Does it frighten you to visit your instructor's office and ask for help?

To help you deal with these difficult situations, I suggest you practice this simple technique. Take about ten minutes to relax and calm yourself down. Do some deep abdominal breathing, or the natural calming breath as described in Chapter 3. Once you are feeling calm, imagine precisely what might occur in the situation and then visualize yourself calmly and peacefully handling it with competence, self-assuredness, and adeptness. See the situation going well for you, and you remain calm and composed throughout. This visualization should be practiced several times in the days before the actual event occurs. It is particularly important to practice it in the half-hour preceding the event.

BUILDING CONFIDENCE

Most students who are successful and confident in math or any other subject have learned the one important thing that sustains them: *focus on the positive.*

Successful students often have a storehouse of positive images in their memory that they can readily recall whenever the going gets rough. Past positive experiences are the foundation for future successful ones. If they recall only positive, pleasant experiences related to math, they boost their confidence whenever they need to.

Is your memory bank a powerhouse of positive images?

Here is an exercise to help you deposit only positive images in your memory bank.

EXERCISE 6-3: A MEMORY BANK OF POSITIVE IMAGES

Each night before going to sleep, think of positive, enjoyable math experiences. Don't be hard on yourself. Evoke only good, pleasurable images. Recall even the smallest accomplishments. Perhaps you remember being proud of yourself as a child for learning the times tables, or when you got the best grade on a math quiz in the seventh grade, or when you did your tax return on your own. Begin to deposit these good math images in your memory. This process will boost your confidence and improve your self-esteem about doing math.

A variation of this exercise is to take a notebook and record all your successes in math. This can help build your sense of accomplishment and self-worth. You need not record major math accomplishments. For example, Dana recorded her correct response when called upon in math class, Marisa wrote that she felt successful every time she balanced her checkbook, and Matthew noted that he finally understood quadratic equations.

Successful students build their confidence on all the little accomplishments accumulated and stored in their memories. Allow the challenge of math to confront all the strength and power of your memory. Whenever necessary, simply draw on this powerhouse of good math memories. There are no penalties for early withdrawal—only lots and lots of interest!

ANCHORING CONFIDENCE

Wouldn't it be wonderful to always be confident and capable when doing math problems or taking a math class? And if at times you weren't, wouldn't you like the power to just snap your fingers and magically feel self-assured and proficient? Well, you can, through anchoring.

Anchoring is a very powerful technique to instantly retrieve past moments when you felt assured, adept, successful, competent, and great about yourself. You can learn to associate these past positive feelings, thoughts, or states of being with any current action or event.

Anchoring is a concept found in the work of Richard Bandler, John Grinder, and other pioneers of Neuro-Linguistic Programming. Using anchoring, you can evoke or visualize a past positive success experience and use this experience now to help you feel more positive about math. Anthony Robbins, in his dynamic book, *Unlimited Power*, notes that anchoring "can create the state you desire in any situation without your having to think about it. When you anchor something effectively enough, it will be there whenever you want it."

Have you ever had the experience of smelling a freshly baked cherry pie, and suddenly you felt transported in time to childhood when your grandmother had baked you a cherry pie? The good feelings of being with your grandmother, or thoughts of your grandmother's house, may return. In this case, the cherry pie acted as your "anchor" to the past, helping you to retrieve the thoughts and feelings of that time in your history. Or perhaps you heard a song you hadn't heard in many years, and it brought on a nostalgic feeling of bygone days. The song acted as your anchor.

Most anchoring occurs without our awareness of the process, but you can consciously use this technique to increase your level of math confidence. Exercise 6-4 will help you do this.

EXERCISE 6-4: ANCHORING CONFIDENCE IN MATH

1. Arrange to be alone in a quiet setting where you will not be disturbed. Sit comfortably with your eyes closed. Begin by breathing slowly and deeply and gradually calm your body and mind.

2. Now, while in a relaxed state, go back over your life to a time when you felt very successful, extremely confident, unbelievably capable and competent. When you evoke this memory, establish a clear mental image of it. Allow yourself to reexperience this time in your history. What were the sights, the smells, the feel, the sounds, and the tastes? How did you look, act, and respond to others? How did others treat you? What thoughts and feelings did you have? Let yourself physically, mentally, and emotionally "feel" the full impact of this experience of total confidence. Hold your head and body the way you did when you felt this way. Notice your shoulders and the position of your spinal column. Breathe the way you did then. At the peak of this reexperienced state, go immediately to step 3.

If you can't identify a time when you felt confident and capable, create your own programmed visualization where you see yourself feeling extremely confident and capable. Again, be sure to experience this image fully, on the physical, mental, and emotional levels. At the peak of this experience, go to step 3.

3. At the peak of feeling confident and capable, snap your fingers (or if you prefer, clap your hands, touch your thumb and index finger together, or do something similar to establish your anchor). By doing this, you are anchoring the positive state of confidence and capability to the snapping of your fingers.

4. Repeat steps 2 and 3 several times in the next few days, with other positive memories or visualizations; each time, get increasingly into the feeling of complete confidence and self-assuredness. At the height of experiencing this strong powerful state of confidence, be sure to anchor the state to your snapping.

5. Soon you'll discover that by just snapping your fingers, you can instantly feel confident and capable. Whenever you are doing math or sitting in math class and you need a confidence boost, snap your fingers! A flood of positive feelings and images will immediately come into your consciousness and help you feel better.

Mary, a pre-med major, used anchoring to help her feel better about taking math. She had been out of school many years and, upon returning, found herself very anxious and insecure in her math class. Following the steps in Exercise 6-4, she learned that snapping her fingers instantaneously brought back a stream of good thoughts associated with competence and success. Mary's anchor (snapping her fingers) helped her reexperience the feelings of being the sixth grade, statewide spelling champion, of earning an A on her seventh grade math final, of receiving an award at high school graduation, and helped her to visualize herself successfully graduating from medical school. Whenever she became fearful or nervous while doing math or when called upon in math class, she'd snap her fingers and a calm sense of success, competence and assuredness would come to her. Backed by these positive feelings, she could think more clearly, figure out her math more easily, and watch her skillfulness increase.

EXERCISE 6-5: VISUALIZING METAPHORS FOR SUCCESS

Assume a comfortable position and close your eyes. Take slow, deep breaths and allow yourself to become increasingly relaxed. In this exercise, I first want you to picture in your mind something about math that makes you feel anxious or uncomfortable. I want you to give life to this image by likening it to an unpleasant but not unfamiliar scene. Here are some examples of images my students have visualized for discomfort and anxiety:

1. Stuck in a deep pit in the ground
2. Feeling claustrophobic in a long, dark tunnel
3. Being in the spotlight in a lineup
4. A wounded animal
5. A tangled, tightly made knot

What images best describe your discomfort?

____

____

____

Now, during your visualization, allow these images to gradually lighten up, fade, soften, and become more relaxing. While visualizing new metaphoric images, repeat reinforcing positive affirmations. For example:

1. While in the deep pit, the sun begins to shine in, and you notice that a ladder is tucked away in the corner. See yourself using the ladder to climb out of the pit. Say the affirmation, "I am able to deal effectively with math."
2. The long, claustrophobic tunnel suddenly opens up onto a sunny bright seashore with a cool breeze. Repeat the affirmation, "Math is becoming easier and easier for me to understand."
3. The lineup spotlight fades into a soft, glowing lantern, illuminating you under the stars on a romantic evening. Say the affirmation, "I am enjoying my math more each day."
4. The wounded animal begins to heal and soon is completely recovered and stronger than before. Repeat the affirmation, "I know I am capable, confident, and competent to do my math."
5. The tight knot is loosened and untangled. State the affirmation, "I allow myself to remain relaxed and calm when doing math."

EXERCISE 6-6: MY BRILLIANT (MATH) CAREER

If someone made a videotape of your life that you could replay on a videotape machine, you would find that your memory of what occurred in the past conflicted with what actually happened.

If you now associate a negative feeling with a past experience, realize two things. First, "negative" is a judgment about an experience, but it is not the experience itself. The judgment is made by your conscious mind. Second, realize that the only thing that happened is that you formed a perception about an event, and that perception now colors your thinking.

Now that you've developed your own ad campaign to promote your new image as a math winner (Exercise 5-5), go one step farther. Become a movie producer and write, direct, produce, and star in the academy award-winning movie, "My Brilliant (Math) Career," which chronicles your life from early successes and triumphs in math to your current state of proficiency and accomplishment. Be sure to dramatize fully the early scenes—in technicolor and Dolby sound—where you shrugged off a reprimand or made light of an embarrassing moment at school and went on to persevere and prove your math prowess to the entire world (fanfare composed by John Williams)!

SUMMARY

With Programmed Positive Visualization, you can program yourself to succeed in math. In this section of our road map, we examined strategies to reprogram negative thinking, to rehearse an upcoming event, to build and anchor confidence in math, and to create a memory bank of positive imagery for math success.

CHAPTER

7

ENHANCING YOUR LEARNING STYLE

As far back as she could remember, Ashley had great difficulty learning math in class, yet recently we found that when she could manipulate things with her hands or if she could move around, math came easily to her. Learning either from the blackboard or from lectures was very frustrating. She had trouble concentrating on math for more than 15 minutes at a time. Her mind would wander and she'd want to do something—anything—but sit in one place. She'd get restless, move around, want to talk to others, or leave the room. As a child, she had often been reprimanded for this behavior regarding math and, early on, she began to develop a strong dislike for the subject.

While working with Ashley, I discovered that she needed to be physically involved in the learning process. Hands-on activities, measuring devices, abacuses, beads, seeds, beans, and other physical objects helped her understand and learn math. She needed to walk or pace back and forth as she studied. This helped clear her mind and increase her concentration so she could understand more. I also found that she needed to munch on carrots or celery sticks as she studied. Munching seemed to keep her energy level and motivation high. She made sure to keep some vegetables on hand, cleaned, and ready to eat. Working under these unique study conditions, Ashley was able to bring up her math grade by a grade and a half.

Jon found that he studied best by himself, late at night, under a bright light in the quiet of his room. Jon could study at other times of the day, but night was when he seemed to get his best thinking done. He noticed that he was definitely more creative then, and he could easily solve even the most difficult of his math problems. Jon also discovered that, when the temperature of his room was cool, perhaps between 67 and 70 degrees, he could function at his top proficiency level. If it were much warmer than that, he became lethargic and sleepy. Following this study routine, Jon learned his math quickly and more effectively.

What type of learner are you? How do you learn math best? How can you set up your learning environment so that you can perform most effectively? This section of our road map will help you enhance your ability to learn math. It is designed to reveal your own personal learning style; that is, the conditions under which you learn best. As you learn math more easily, you will notice that many of your math fears will gradually decrease.

Would you like to know what you can do to improve your understanding and retention of new and difficult math concepts? Would you like to discover what positively affects your powers of concentration? By gaining an awareness of your personal learning style you'll be able to improve in these areas.

Your learning style has been forged from a unique mixture of personal attributes and preferences, individual background, childhood experiences, and environmental cues of all kinds. By knowing how these factors influence you, you'll be able to intentionally choose the most effective learning environment and strategies to meet your individual needs.

Learning styles can make a big difference in your life. By understanding and working with your unique learning style, you can greatly enhance your math achievement. You'll study better, feel more excited about learning math, and your test scores will be higher. And, what's more, you'll feel a greater measure of self-control.

PERCEPTUAL LEARNING CHANNELS

Your perceptual preferences may be one of the most significant factors influencing your ability to learn and recall math. Three major perceptual learning channels have been identified: visual, auditory, and kinesthetic/tactile. People who are visual learners learn best by *seeing* or visualizing words and numbers written out. Auditory learners generally learn best through *hearing* math explained to them or saying math to themselves. Kinesthetic/tactile learners need to be involved in the learning process through *touch* even *whole-body movement,* if possible. You may find that one of these channels is dominant for you and that a second one will further strengthen your learning.

Professor Rita Dunn, Director of the Center for the Study and Teaching of Learning Styles at St. John's University in New York, has found that when students are introduced to new material through the perceptual channel they prefer most, they remember significantly more than when they are taught through their least preferred channel. In addition, if the new material is further reinforced through secondary or tertiary perceptual preferences, they achieve even more.

Here is an exercise designed to help you determine which perceptual channels you prefer most for learning math.

EXERCISE 7-1: ASSESSING YOUR PERCEPTUAL LEARNING CHANNELS

Carefully read the sentences in each of the following three sections and note if the items apply to you. Give yourself three points if the item usually applies, two points if it sometimes applies, and one point if it rarely applies.

Are You a Visual Learner?

1. ___ I am more likely to remember math if I write it down.
2. ___ I prefer to study math in a quiet place.
3. ___ It's hard for me to understand math when someone explains it without writing it down.
4. ___ It helps when I can picture working a problem out in my mind.
5. ___ I enjoying writing down as much as I can in math.
6. ___ I need to write down all the solutions and formulas in order to remember them.

7.____ When taking a math test, I can often see in my mind the page in my notes or in the text where the explanations or answers are located.

8.____ I get easily distracted or have difficulty understanding in math class when there is talking or noise.

9.____ Looking at my math teacher when he or she is lecturing helps me to stay focused.

10.____ If I'm asked to do a math problem, I have to see it in my mind's eye to understand what is being asked of me.

________ TOTAL SCORE

Are You a Kinesthetic/Tactile Learner?

1.____ I learn best in math when I just get in and do something with my hands.

2.____ I learn and study math better when I can pace the floor, shift positions a lot, or rock.

3.____ I learn math best when I can manipulate it, touch it, or use hands-on examples.

4.____ I usually can't verbally explain how I solved a math problem.

5.____ I can't just be shown how to do a problem; I must do it myself so I can learn.

6.____ I've always liked using my fingers and anything else I could manipulate to figure out my math.

7.____ I need to take lots of breaks and move around when I study math.

8.____ I prefer to use my intuition to solve math problems, to feel or sense what's right.

9.____ I enjoy figuring out math games and math puzzles when I learn math.

10.____ I learn math best if I can practice it in real-life experiences.

________ TOTAL SCORE

Are You an Auditory Learner?

1. ___ I learn best from a lecture and worst from the blackboard or the textbook.

2. ___ I hate taking notes; I prefer just to listen to lectures.

3. ___ I have difficulty following written solutions on the blackboard, unless the teacher verbally explains all the steps.

4. ___ I can remember more of what is said to me than what I see with my eyes.

5. ___ The more people explain math to me, the faster I learn it.

6. ___ I don't like reading explanations in my math book; I'd rather have someone explain the new material to me.

7. ___ I tire easily when reading math, though my eyes are OK.

8. ___ I wish my math teachers would lecture more and write less on the blackboard.

9. ___ I repeat the numbers to myself when mentally working out math problems.

10. ___ I can work a math problem out more easily if I talk myself through the problem as I solve it.

________ TOTAL SCORE

My dominant perceptual learning channel is:

____

(enter the category with the highest total score)

My secondary perceptual learning channel is:

____

(enter the category with the second highest score)

My tertiary perceptual learning channel is

____

(enter the category with the third highest score)

SUGGESTIONS FOR DISTINCTIVE LEARNING CHANNELS

VISUAL LEARNERS

Are you a strong visual learner? Do you find that you must see math problems written on the blackboard or on paper before you can begin to understand and comprehend what is being asked of you? Would it drive you crazy if you had to listen to a math lecture and you had nothing to write with or if the teacher wrote nothing on the board?

Ted, a construction worker, returned to college after being out of school for almost ten years. He really wanted to get his college degree, but math was terribly frustrating to him. He could easily copy everything his instructor put on the blackboard, but he was completely lost with the lecture. Math just didn't make sense to him when he listened to it. Ted learned that he is a strong visual learner but a weak auditory learner. Once he understood *how* he learned, he began using strategies to help him gain the most out of his math classes and his studying. Soon Ted's math achievement and his enjoyment of math began to improve. Ted has continued to take math courses and is now doing well in calculus and loving it.

Here are strategies that may be helpful to you if you are a strong visual learner but are weaker in the auditory channel.

1. Always take written notes when someone is explaining math to you.
2. Whenever possible, ask for written instructions.
3. Make your own drawings or diagrams when figuring out word problems.
4. Use flashcards to review all important concepts, formulas, theorems, equations, and explanations.
5. Write as much as you can when you study. Work out lots of problems.
6. In lecture, concentrate on what the instructor is writing on the blackboard and copy everything down. If you can't get much of what the teacher explains in class, bring a tape recorder. Always reset the counter on the recorder to number

one at the beginning of the lecture. At points in the lecture where you don't fully understand what the teacher says, note the counter number on the recorder and put it down in your notes. Later, you can listen carefully to the tape, paying special attention to the sections where you jotted down the counter numbers. Write down the information from the tape that you missed getting the first time.

7. Use two or more math books. Read how different authors explain the topics you learn. Because this is such a good study skills technique in general, I will discuss it in further detail in Chapter 8.
8. Visualize in your mind's eye the math concepts you are learning.
9. Use computer programs that illustrate concepts you are learning.
10. Read your textbook assignment and previous class notes before your next class.
11. Use workbooks, supplemental study guides, handouts.
12. Map out, chart, or in some way graphically illustrate your classroom and textbook notes.
13. Always write in your textbook. Underline key words. Mark important concepts and use colored pencils to liven them up.
14. Sit near the front of your classroom to avoid visual distractions and to pay closer attention to your instructor.
15. When you review your classroom notes, creatively highlight the important points with colored pencils or markers.

AUDITORY LEARNERS

Are you primarily an auditory learner? Do you prefer to have someone explain math to you rather than read about it or see it on paper? Do you often have to repeat math problems aloud or in your head before you can figure them out? Do you just hate it when a teacher shows the class how to figure out a math problem on the board, but doesn't explain each step aloud while writing it?

The following suggestions will be particularly helpful if you are a strong auditory learner but are weaker in the visual area.

1. Sit near the front of the classroom so you can clearly hear your teacher without auditory distractions.
2. You may want to use a tape recorder during lectures and listen to each lecture as soon after class as possible. Listen to it over and over again, when you drive, study, jog, or do your chores.
3. Take part in classroom discussions.
4. Ask lots of questions in class, after class, and in help sessions. Ask for clarification if you don't completely follow an explanation in class.
5. Restate, in your own words, math concepts you are trying to understand.
6. Ask your math teacher to repeat important concepts.
7. Listen carefully to the math lecture. Mentally follow the concepts, then write them down to capture what was said.
8. If you can't get everything that the teacher writes on the blackboard, find a classmate who seems to be more of a visual learner and is writing everything from the board. Ask if you could photocopy this person's notes after class.
9. When figuring out a difficult homework assignment, you may want to read it aloud into a tape recorder and then listen to it and write it down.
10. Immediately after you read your math textbook assignment, recite aloud what you have just learned.
11. Read your class notes and textbook notes aloud. Whenever possible, say them in your own words into a tape recorder.
12. Talk about math to a study partner or to anyone who might listen. (I know some students who have even explained their assignments to their pets.)
13. Listen for key words in your math lecture. Note if your instructor emphasizes certain points through his or her tone of voice, emphasis on certain words, voice inflections, and so on.
14. Record all the key concepts, formulas, explanations, and theorems on an audiocassette and listen to them often.

KINESTHETIC/TACTILE LEARNERS

Was your score on the perceptual learning channel assessment highest in the kinesthetic/tactile area? Do you prefer real-life experiences with math, manipulating it, and experimenting with it? Do you find that you like to move around when you study, pace the floor, or shift positions a lot?

Here are some strategies that may be useful to you if you are a kinesthetic/tactile learner.

1. You must use a hands-on approach to learning. Work out as many math problems as possible. Do, do, do. Practice, practice, practice. You'll be amazed at the positive results.
2. Whenever possible, convert what you are learning in math to real-life, concrete experiences. If applicable, use measuring cups, measuring vials, toothpicks, seeds, stones, marbles, paper clips, rulers, sticks.
3. If someone shows you how to do a problem, immediately ask if you could work out a similar one to see if you understand how to do it.
4. While studying, try to solve problems several different ways in order to decide which method feels right to you.
5. Many kinesthetic/tactile learners find that they must move during the learning process. You may want to walk to and fro while reading your assignment or even while working out problems. Some students like to rock back and forth. Others need to shift positions frequently. The movement seems to increase understanding and comprehension for some highly kinesthetic people.
6. Use computers and workbooks.
7. While you exercise or engage in other types of physical activities, review your math concepts in your mind.
8. Use your fingers and even your toes if this helps when you figure out math problems.
9. Rewrite class notes.
10. Use a calculator to solve problems.
11. If possible, use or build models to help you understand math concepts you learn.

"JOAN'S A KINESTHETIC LEARNER"

12. Study math on an exercise bike—preferably one that has a reading stand attached to it and that allows you to move your arms as well as your legs.

STRENGTHEN ALL AREAS

I recommend that you strengthen and connect all your perceptual learning channels. For example, recite aloud what a diagram or chart represents so that you both see and hear the information. You may also decide to redraw it or to write out an explanation of it so you can be kinesthetically involved. Each channel can reinforce the other. This is particularly helpful because you may not have the luxury of being in a course that readily fits your preferred learning style. I believe that if you strengthen all modes of learning, you'll be better able to process information on both sides of your brain. Then, if you don't recall the information you've learned one way, you may be able to recall it via another. So see the information, say it, hear it, write it down, feel it, visualize it, read it, and manipulate or work with it in as many ways as you can.

OTHER MAJOR FACTORS AFFECTING YOUR LEARNING

We are all biological creatures affected by a bewildering assortment of stimuli, both internal and external. We get hungry and we eat; we get full and sleepy; we are startled by noise or lulled by it; we fidget with energy or drowse in repose. Our attention is diverted by so many distractions, it's sometimes hard to imagine getting our brains to focus on learning new material. But learn we must. And, if we want to do that most efficiently, we need to figure out how to analyze these influences and either minimize the disturbance they cause or maximize their positive effects.

TIME OF DAY

When is your energy at it's highest? Are you an early bird ready to go with eagerness and vigor at six or seven in the morning? Or are you a night owl like Jon, who was mentioned earlier in this chapter? Maybe you feel as if you're in a fog until 10 A.M. and that it would be a crime to force you to take an 8 or 9 A.M. math class. Perhaps you become a shining star at noon or in the early afternoon. Or, possibly, you are roaring your engines at 7 P.M. and can go full steam ahead until midnight.

Lanny, a student in my math anxiety reduction course, said she became pretty discouraged with math in high school. She was always scheduled to take math first thing in the morning, usually a 7:40 A.M. class. She found that she could never concentrate in class and couldn't grasp the concepts. She began to hate math. Now, when she looks back at her high school years, she realizes that she was a confirmed night owl, staying up late each night and never really coming life the next day until around 11 o'clock in the morning. Lanny is still a night owl, but now she makes sure she schedules her math classes in the afternoon or evening, and she never takes morning classes. She studies best after everyone in her family is sleeping soundly.

Each of us has times in the day or evening when we perform at our peak level of proficiency. If you schedule your math classes or math study sessions at these times, you'll find that you can concentrate better and learn more. Exercise 7-2 will help you determine the hours of your peak performance.

EXERCISE 7-2: WHEN IS YOUR MENTAL ENERGY AT ITS HIGHEST?

Notice your mental energy level throughout the day for a few days. On the following chart, use one check to indicate the times you feel alert, clear, and can concentrate and study well. Use two checks when you function at your peak: think best, learn more quickly, concentrate deeply, understand clearly.

SOUND LEVEL

Do you like to study in a very quiet place free from all distractions? For most people this seems to help increase comprehension and the ability to figure out difficult problems.

Francine preferred silence and was easily disturbed by any sounds, particularly when she was nervous about an upcoming math test. Francine found that wearing small moldable ear plugs, which she bought at a local pharmacy, did the trick. She wore them every time she studied at home or in the library. She found that her concentration and retention levels increased markedly.

	DAY 1	DAY 2	DAY 3	DAY 4	DAY 5
5 AM					
6 AM					
7 AM					
8 AM					
9 AM					
10 AM					
11 AM					
NOON					
1 PM					
2 PM					
3 PM					
4 PM					
5 PM					
6 PM					
7 PM					
8 PM					
9 PM					
10 PM					
11 PM					
MIDNIGHT					
1 AM					
2 AM					

Many students find silence especially helpful when they work on demanding or problematic assignments, but they seem to tolerate some background music or noise when they do a routine or boring assignment or when they recopy notes.

Other students find that, when it's too quiet, they become "hyper-aware" of all sounds. Howard, an accounting major, always studies with the television or radio on in the next room, although he doesn't listen to it. He says that if it's too quiet, he hears the refrigerator motor, the ticking of the living room clock, and the sounds of his own heartbeat.

Harriet, a statistics student, found that playing soft, inspiring music increased her motivation for studying and the amount of material she covered. Tony, a calculus student, learns best when listening to music of classical baroque composers. The slow tempo of approximately one beat per second calms him and clears his mind.

EXERCISE 7-3: HOW DOES SOUND AFFECT YOU? (*optional*)

Briefly describe which sound conditions work best for you when you study math.

LIGHTING

How does lighting affect your ability to study and learn? Do you find that you study best with natural light coming in from the window? Or perhaps you like to cuddle under a soft, incandescent lamp. Perhaps you function best under white fluorescent bulbs or a bright halogen light.

Some students find that bright lights energize them and make them more alert and attentive. When the light is subdued, they lose interest easily and become distracted, apathetic, and drowsy. Other students find bright lighting has the opposite effect on them; it makes them tense, fidgety, and uncomfortable.

Many people feel more relaxed and concentrate better when they work in diffused natural light or under balanced full spectrum fluorescent lights. Experiment with various lighting conditions and figure out for yourself how lighting affects you.

EXERCISE 7-4: HOW DOES LIGHTING AFFECT YOU? (*optional*)

Briefly describe what lighting conditions work best for you when you study.

TEMPERATURE

Do you prefer to learn in a room that is cool or one that is moderately warm? Some students think better when the room temperature is cold, whereas others find that this is intolerable and distressing.

Each of us has our own unique reaction to temperature. What if you're taking math in a warm classroom and you think better in the cold? Why not wear very lightweight clothes to class? Consider sitting near a window or an open door where you might get some cooler air, and be sure not to sit under heating ducts. If the opposite situation occurs and the room is too cold, take extra layers of clothing to class to keep yourself warm. You also might notice that, at different times of the year or at different times of day, your reaction to environmental temperature varies. If you're extra tired, catching a cold, or disturbed by a personal problem, you might find yourself colder than usual.

EXERCISE 7-5: HOW DOES TEMPERATURE AFFECT YOU? (*optional*)

Briefly describe what environmental temperature conditions work best for you when you study math.

ROOM DESIGN

Do you like to study in a big, soft, comfortable chair or couch? Or perhaps you study best at a desk seated in a nicely cushioned straight-backed chair. Perhaps you like to sprawl out on a carpeted floor with all of your papers and books spread out around you and soft fluffy pillows to lean on. For some students, this eases their tension and motivates them to study more.

Andrea found that if she sat in a comfortable armchair she soon would become sleepy and her eyes would close. She would become too relaxed and she'd rapidly lose all motivation for studying. To do her best, Andrea arranged to study at a large uncluttered desk where she could use the entire top surface to spread out her math books, assignments, class notes, scratch papers, and practice worksheets. This arrangement helped her to see everything clearly and to stay in a problem-solving mode. At school, she found that if she sat in a study cubicle in a quiet corner of the library, she stayed right on target. She avoided the large comfortable library armchairs where she saw some of her classmates snoozing.

When you prepare for tests, I encourage you to take all your practice exams in a room arranged much like the testing room. So, although you might prefer to learn in a comfortable setting, at a large desk, or in the middle of your bed, you must be able to transfer this learning to the classroom and testing situation. This transfer will occur more easily if you give yourself practice exams in a room of similar design.

EXERCISE 7-6: HOW DOES ROOM DESIGN AFFECT YOU? (*optional*)

Briefly describe in what setting you learn and concentrate best.

FOOD INTAKE

Do you often have the "munchies" when you study, and you can't concentrate unless you have something in your mouth? Or perhaps you find that you must wait at least an hour or two after you eat before you can concentrate on learning anything.

You may have noticed that certain foods affect your study capabilities more than others. Too much caffeine may make you nervous and jittery, and a large fatty meal may dull your senses and make you feel exhausted and lethargic. A few ounces of protein eaten an hour or two before you study may help increase your alertness and motivation. An ounce or two of carbohydrate eaten before or while you study can calm you and increase your focus. Chapter 9 discusses some of the effects of food on the thinking process.

EXERCISE 7-7: HOW DOES FOOD AFFECT YOUR STUDYING? (*optional*)

1. Do you think better if you snack when you study?
2. What foods seem to affect you positively?
3. What foods negatively affect your ability to study or make you lethargic and tired?
4. How does each of your meals affect your ability to concentrate?
5. How long after you eat a meal do you reach your best level of concentration?

In the space below, briefly describe how food affects your ability to study math.

ALONE OR WITH OTHERS

Are you a loner when you study? Do you like to review and study with a partner or in a small group? Perhaps you enjoy studying in a math lab where you can be alone but where there are others around who are studying math and where tutors available if you need help.

CLOTHING

Do you prefer to wear very loose clothing when you learn math, or is this not an issue for you? In class and when taking tests, make sure you wear clothes that are as comfortable as the clothes in which you study. What are your favorite clothes for studying? What clothes make you feel successful and sure of yourself?

PERSONAL THINKING STYLES

From the time he was quite young, my son's style, when given a new electronic device, was (and still is, at age 23) to start to use it immediately, before reading the directions. This approach rapidly solidified his functional grasp of that machine's operation. Written materials were used only as a last resort ("when all else fails, read the instructions").

Contrast this with the style of the person who will not touch the gadget before mastering all written directions, methodically working through them step by step, and who actually may be intimidated by its mechanical aspects ("if all else fails, plug it in").

In between are mixtures of the two extremes—those who combine a commonsense willingness to follow written instructions with a spirit of adventure and a readiness to derive flashes of insight from a trial-and-error approach.

When it comes to learning and applying mathematics, where do you fit into this broad analogy? Is your personal thinking style to immediately begin to work out and solve problems, allowing the insights you gain by doing to give rise to the broader generalizations and principles that apply to all similar problems (inductive process)? Or do you prefer to assimilate theory and background until you have a clear concept of the principles involved, and only then apply them to individual problems one at a time (deductive process)?

DOERS

Are you an inductive "doer"? If so, you may prefer to do math problems before you've understood your assignment or the principles involved. If this characterizes you, I suggest that when you are presented with a new math topic, work out lots of problems in order to get a good feel for that particular concept. Attempt to identify the commonalities among the problems so that you can ascertain the unifying principles involved and later apply these principles to other more difficult problems.

Each time you figure out a new principle, test it out to see if it really works. See how many different ways the new principle you've learned can actually be used in real-life examples. This is an excellent approach to enhance your learning, but at times even this approach may not be adequate. For example, many doers experience anxiety when faced with problems for which the solutions require new knowledge and not just number crunching, thus underscoring their lack of information. If this happens frequently to you, it would be helpful for you to go back to basics, recognize the deficiency, and develop the good study-skills habit of reading the assigned materials first before attempting to do homework problems.

COGITATORS

Perhaps you're a deductive "cogitator." If you are, you may prefer to first grasp theory and principle before attempting to do any of the assigned problems. This is an excellent approach, since it's certainly easier to memorize math rules if you understand them first. But math concepts are often hard to fathom at the onset, and this may cause you to get stymied and experience anxiety or distress. If this happens to you, try easing into problem solving like sliding into a cool stream, gently testing the water. Sometimes only by working out problems do certain principles begin to make sense. First, work through sample problems in your textbook that show the solutions step by step. Do as many sample problems as you can from your text as well as from other textbooks and college course outline series. From there, you can move on to problems of progressively increasing difficulty, always doing only problems for which the book offers the solution. Keep going back and forth between reading the principles and seeing how the principles are put into action. This process will help you to better understand the concepts and principles you're learning. Whenever you feel stymied by a new concept, you need to start to put the principles into practice. Soon you'll feel much more comfortable.

SUMMARY

You can greatly enhance your ability to learn math if you remain aware of, and at times also manipulate, the factors that influence your unique learning style. In this section of our road map, we assessed whether you are a visual, auditory, or a kinesthetic/tactile learner, and I offered you learning strategies specifically adapted to your preferred perceptual learning style. We also examined how you are affected by an assortment of internal and external stimuli such as time of day, sound, lighting, temperature, food intake, social environment, and clothing. We concluded the chapter with a discussion of two major personal thinking styles: inductive (doers) and deductive (cogitators).

CHAPTER

8

EFFECTIVE MATH STUDY SKILLS: THE FUEL OF EXCELLENCE

Imagine for a moment that you're giving a surprise party for someone special. You've invited people you care about, and you want everything to go perfectly. Are the balloons inflated? Does the party table look just right? Are food and drinks set out? Are the flowers fresh? Has just the right music been selected? Don't you think things are more likely to go the way you've planned them if these arrangements are in order? Preparations pay off. And if you want your journey toward your math goal to go smoothly, with less hassle, it's absolutely crucial that you set the stage for learning. How? By sharpening your math study skills.

Don't shrug your shoulders. Math isn't quite the same as other subjects. Strategies that help you study and learn psychology, history, or English won't necessarily help you with math. There are studying techniques that work best for math; master them, and master good habits that breed success. Master them, and you will remember math concepts more easily, retain them longer, and use them more effectively.

This chapter explores these study-skills methods in detail. I urge you to approach them seriously. They have already provided many math anxious students with major breakthroughs on their road to math success. Why not you? At times, these skills may seem disarmingly simple, and you may recognize and already know about some of them. But, believe me, they are more powerful than you think! They

are choice ingredients in your math mastery recipe. And remember, using good study techniques won't necessarily mean you have to study more hours; it means you will study more efficiently and obtain gratifying results.

HOW GOOD ARE YOUR STUDY SKILLS?

EXERCISE 8-1: MY PERSONAL MATH STUDY SKILLS INVENTORY

This inventory will help you assess the effectiveness of your math study skills. Read each of the statements carefully and determine how frequently each applies to you. Enter the correct point score for that item (usually = 3, sometimes = 2, and rarely = 1).

	Usually (3 points)	Sometimes (2 points)	Rarely (1 point)
1. I attend all my math classes.	_____	_____	_____
2. I read my math assignment before attending class.	_____	_____	_____

3. In class, I mentally follow all explanations, trying to understand concepts and principles. _____ _____ _____
4. In class, I write down main points, steps in explanations, definitions, examples, solutions, and proofs. _____ _____ _____
5. I review my class notes as soon after class as possible. _____ _____ _____
6. I review my class notes again six to eight hours later, or definitely the same day. _____ _____ _____
7. I do weekly and monthly reviews of all my class and textbook notes. _____ _____ _____
8. In reviewing, I use all methods, such as reciting aloud, writing, picturing the material, etc. _____ _____ _____
9. I study math before other subjects, and when I'm most alert. _____ _____ _____
10. I take small breaks every 20 to 40 minutes when I study math. _____ _____ _____
11. I work to complete my difficult math assignments in several small blocks of time. _____ _____ _____
12. I reward myself for having studied and concentrated. _____ _____ _____
13. I survey my assigned math readings before I tackle them in depth. _____ _____ _____
14. When I read, I say aloud and write out important points. _____ _____ _____
15. I underline, outline, or label the key procedures, concepts, and formulas in my text. _____ _____ _____
16. I take notes on my text and review them often. _____ _____ _____

17. I complete all assignments and keep up with my math class.	_____	_____	_____
18. I study math two hours per day, at least five days a week.	_____	_____	_____
19. I work on at least ten new problems and five review problems during each study session.	_____	_____	_____
20. I work to "overlearn" and thoroughly master my material.	_____	_____	_____
21. I retest myself often to fix ideas in memory.	_____	_____	_____
22. I work to understand all formulas, terms, rules, and principles before I memorize them.	_____	_____	_____
23. I use a variety of checking procedures when solving math problems.	_____	_____	_____
24. I study with two or more different math books.	_____	_____	_____
TOTALS:	_____	_____	_____

To find your total score, add up the total points of all three columns:

MY GRAND TOTAL IS: _____________

If your score is above 68 points, you have excellent math study skills.

If your score is between 54 and 68 points, you have fair study skills, but you need to improve.

If your score is below 54 points, you have poor math study skills and you need help fast!

Now let's see what you can do to improve your math study skills.

HOW DO YOU APPROACH LEARNING?

One traditional approach to the learning process is the *empty vessel* model. In this model, the student considers his or her mind to be an empty container that the teacher is expected to fill with knowledge. If as a student you adhere to this approach, you would, for example, register for several classes; you would then expect to learn all there is to learn by simply attending class and opening your mind to the subject matter, which then pours in. I am *not* a believer in this approach. My many years as a college counselor have shown me that learning just doesn't happen this passively.

We are very complicated beings, and we must understand ourselves and how we learn best before we can reach out and tackle a subject. We must understand how we think, create, and retain new information. We must understand the memory process and strategies that enhance recall. We must look at the uniqueness of our subject matter and identify how best to approach each new area of knowledge. This is the *personal growth* model of learning. The user-friendly strategies described here are based on this approach.

SELECTING A MATH CLASS

Let's begin our discussion of math study skills with your plan to take a math class. This is a very important step, and you must execute it carefully. There are four aspects you should consider: (1) correct placement, (2) when you last took math, (3) auditing or repeating a math course, and (4) selecting a good teacher.

Correct placement in math is crucial. If you are put into a math course that is too advanced for your level of understanding, your anxiety level is sure to increase. Find out if your school has a math placement exam. I encourage you to sit down with a math teacher who can help you assess your math skills and place you properly.

When did you last take math? I encourage people who are taking math never to skip a semester if they can help it, because it's easy to forget a lot fairly quickly. One study found that after one year of nonuse, students had lost approximately two-thirds of their elementary algebraic knowledge (as quoted in Pauk, 1974). If you've been out of school for many years, you are sure to have forgotten much more of your math. A math placement exam could help you decide which math course to take.

As a rule of thumb, if you haven't taken math in about three years, you probably will need to repeat the last math course you successfully completed. And think of successful completion as a grade C or better. In fact, recent findings have indicated that students are more likely to succeed as they go into higher and higher levels of math if they have no lower than a B in their algebra courses.

If you decide to repeat a course for credit and you receive a higher grade than you received before, most colleges will use only the higher grade in your grade point average (GPA) and eliminate the lower grade from your GPA. Find out how your school handles repeats.

If, in order to freshen up your skills, you decide to repeat a course you have already taken, you may consider auditing the course so you don't have to be concerned about the grade. But I urge you when auditing the course to come to class regularly, complete all the homework, and take all the tests. You will get the most out of the course this way, without having to worry about a grade. Auditing can be used in another way as well. If you plan to take a math course that you feel is a difficult one, you might audit it first and take it for credit the following semester. The audit allows you to familiarize yourself with the material, and it gives you the opportunity to practice working out the math problems. Once again, I strongly encourage you to attend all classes, do the homework, and take the tests, even though you will not get a grade. Don't audit unless you plan to participate.

TEACHER SELECTION: SATISFACTION GUARANTEED

Once you have decided which math course you will register for, the next critical step is to choose a good math teacher. Spend time asking around about math teachers. Ask others why they like or dislike a certain teacher. Don't choose a class just because it fits into your schedule well.

A great teacher can make all the difference in how you feel about going to class and how much you learn in class. Select a teacher who explains concepts well; teaches according to your learning style (see Chapter 7); welcomes questions before, during, and after class; has office hours for outside help; has a positive attitude toward students; and gives fair tests.

We have all had teachers whom we remember fondly (or not so fondly) and whose image and manner are inexorably linked in our minds with the subject they taught. Some students actually are so put off by a teacher's appearance or personality that they come to dislike the course and, by association, the subject matter also. By "losing the forest for the trees," they sabotage their progress.

You may also want to know if your teacher teaches more than one section of the math course you plan to take; if you have to miss a class or if you haven't quite gotten a certain concept in class, you could ask to sit in another section of the same course at a different time.

GIVE YOURSELF THE TIME

Once you've chosen the appropriate course and the best teacher for you, you are ready to attend a math class. I encourage you to view the learning of math as a positive and rewarding academic challenge. Devote your energy to it. It requires persistence, concentration, discipline, patience, and lots and lots of practice. Don't take math with other hard courses or a busy work load. Give yourself the time! Many teachers and students have learned the "rule" that you should study two hours for every one hour you are in class. But this may have no basis in reality when it comes to math or any other difficult subject. Successful math students usually study math for at least two hours every day throughout the semester. So don't feel bad if you don't assimilate math as fast as you think you should. Learning math takes time. Give yourself the time; nothing succeeds like excess!

STAY CURRENT

It is most important to stay current in math. Don't fall behind or the entire course will become an effort and a struggle for you. Success in math builds on existing knowledge at each stage. Be attuned to the cumulative nature of math. You can only understand new information if you assimilate and digest earlier information. So keep up with the work and don't fall behind; try not to miss important building blocks along the way.

ATTEND ALL CLASSES

Successful students are more likely to attend all classes, whereas failing students miss one-third or more of their classes. Don't cut math class. Missing even one class may actually put you behind by at least two sessions, because you may feel lost when you return to class the next session. If you have to miss, be sure to read the assigned text material thoroughly, do your homework, and get a copy of the class notes from a classmate. Go to your teacher to clear up anything you don't understand. If you do this all before attending the next period after your absence, you will be in a better position to understand and follow the class lecture. Your reward for persistence is peace of mind.

BE BOLD: SIT NEAR THE FRONT

Successful students are also more likely to sit close to the front of the classroom and near the center. Be bold—boldness has genius. You are more likely to pay attention and concentrate on the lecture by sitting close to the front. It also helps you to be more involved in the class, to have more direct contact with the teacher, and to see the board more easily. Those students who sit in the back of the class are not only physically but also psychologically more distant from the lecturer. It is much easier to get distracted by sounds or side discussions going on in the back of the room.

TAKE FULL CLASS NOTES

Your class notes and your text notes are like your bible in math. They indicate the essence of what you are learning. Studies show that successful students take fuller class notes—about 64% of what is presented—than unsuccessful ones. So take complete notes in math class. Write down what the teacher puts on the blackboard and all verbal explanations that can clarify what you are learning. Write down important ideas, equations, examples, helpful hints, and suggestions. Strive to follow and understand the teacher's reasoning and logic when solving a board problem. Note steps in a solution that the instructor explains but doesn't necessarily write on the board. Even if you become confused when you listen in class and don't fully grasp the ideas being presented, keep taking notes. Don't be afraid to ask the instructor to repeat anything you miss or don't understand. If you document what has gone on in class, you can refer to it later and work to increase your comprehension. If your teacher explains examples directly from your math text, you needn't write these down. Simply follow along in your text and add any clarifying statements. Make your notes legible, neat, and clear so you can read them easily.

Be an active listener. No teacher can speak or write as fast as we can think, so it is very easy to "tune-out" during math class. I hear this from lots of students who have trouble concentrating in math class. They begin to daydream, or think about the chores they need to do or about a relationship they're in, and pretty soon they've lost the train of thought in the lecture. So be an active listener even when gaps occur in class. For example, relate what is being taught to previous lectures, to the homework assigned, or to your textbook reading. Pose questions that can keep up your interest and further your understanding. If you've read the text prior to the lecture, you should jot down points that need clarification. Ask about them if the instructor doesn't clear them up.

Choose a large notebook with pockets that you can use exclusively for math. In the front half of your notebook, write class notes, and in the back half write textbook notes and the solutions for homework and sample practice problems. Use the pockets for handouts, syllabuses, or returned test papers. Date each day's class notes and identify the topic on top of the page in large writing. If you like being artistic, write topic headings with colorful markers. Leave plenty of space in your class notes for additional clarifications, diagrams, sketches, and comments you may want to add later. It is also best to write only on one side of the page so that you can use the reverse side for comments, additional sample problems, or questions about areas of uncertainty.

It's also an excellent idea to use your colored markers when you review your class notes to identify definitions, theorems, proofs, formulas, procedure steps, examples, or equations. If you clearly label your class notes, you will be more efficient in locating information for review and study for tests.

QUESTIONS THAT COUNT; ANSWERS THAT ADD UP

Always remember, you have the right to ask questions of your teacher before, during, and after class. See your instructor during office hours or visit the math learning center, if your school has one. Notice when you begin to get into trouble and seek help immediately. Never avoid asking questions because you are afraid to look stupid. There is *no such thing as a stupid question!* All questions are relevant because they help clarify difficult steps in procedures, and increase your understanding. So ask questions and talk your way to success!

Be creative. Design questions for yourself when you read your chapter, when you do your homework assignments, and when the teacher is explaining concepts in class. If you have difficulty following a procedure your instructor is working out, ask questions about each step that you find confusing. Never allow your questions to go unanswered; reach out, take the initiative, and get help from your teacher, a tutor, or a friend. Be resourceful and trust your creativity—it is your hidden power!

CAPITALIZE ON THE MAGIC OF "NOW"

I've had many students who come for counseling and tell me that they understood the material presented in their math class, but when they looked at their notes and homework a day or two later, it was all "Greek." They must then sit for hours reconstructing what it was that their instructor taught a day or two earlier in class. If this process continues, they soon begin to fall further and further behind. It's a vicious downward spiral, until they feel that they are too far behind to catch up and then consider dropping the course.

I have found that the most important study skill that math students can learn is to *review immediately after learning* and then again eight hours later. This review directly after math class is critical.

"PARDON ME SIR - COULD YOU PLEASE EXPLAIN STEP 3?"

ASK QUESTIONS

Reviews need only last 10 or 15 minutes, since you already know the material. You may have to fill in words, numbers, or symbols you left out because you were writing so fast in class. Cover your notes and repeat them to yourself in your own words or picture them in your mind. Cover up the solutions to problems worked out in class, and see if you can work them out now. It is also important to do your homework the same day as your teacher assigns it. This acts as a review of what you just learned. Then do weekly and monthly reviews of your notes to get the information firmly embedded in your memory. Make a commitment to yourself to review regularly throughout the semester.

Let me explain to you why it is so important to review immediately after learning. Almost immediately, you lose information you first learn in math class or from reading a math book. The German psychologist Hermann Ebbinghaus was the first person to do research on the rate of forgetting. He did his research with meaningless material such as nonsense syllables. He found that, after 20 minutes, nearly half of what had been learned was forgotten, and after one day nearly two-thirds was lost. Ebbinghaus also found that after 2 days, 69% was lost; in 15 days, 75%; and in 31 days, 78%. That means that after a month, you remember only 22% of the material you learned.

A classic study by H. F. Spitzer on the retention of meaningful material found results similar to Ebbinghaus's study (as quoted in Pauk, 1974). Spitzer concluded that even with meaningful material, most forgetting takes place immediately after learning occurs. However, Spitzer also found that students who reviewed the material immediately after learning and then did periodic reviews were able to retain almost *80%* of the material after two months! Why this astonishing difference?

One reason that reviewing immediately after learning seems to be so effective is that it takes time for the temporary memory to be consolidated or converted into permanent or more long-term memory. There is a reverberation circuit of neurons in the brain that helps in this process. If you stay with the material you just digested and review it immediately after learning, this circuit is repeatedly activated until your memory or *neural trace* is strengthened, forming long-term memory. Estimates indicate that it takes between 4 seconds and 15 minutes for a memory trace to consolidate or jell in your mind. If you don't review immediately after learning new material but instead jump to another subject, watch the news, or run off to another class (as with back to back classes), you prevent this memory trace from becoming established in your long-term memory. So stay with it, review soon after learning and often thereafter. You'll find that you will start to remember more. And, because you know the material, you don't have to spend hours relearning it. This is the most efficient and effective way to learn math. So don't delay; review your material promptly and capitalize on the magic of *now!*

THE BIRTH OF EXCELLENCE: SAY AND DO

In the personal growth approach, I encourage you to be fully involved in the math learning process. This means use all your senses: recite the material aloud; explain it to others; hear it; see it; write it down; work with it; manipulate it in as many ways as possible; work out as many problems as you can. This ensures that, if you can't remember your material one way, you will another. If you use a variety of methods you will also be reviewing your material and reinforcing your memory of it. This approach to learning is further substantiated by memory retention studies (as reported by Magnesen, 1983) showing that students tend to remember a full 90% of what they "*do* and *say*."

So jump right in and be involved! Make sure to take both class notes and reading notes. Taking notes is an active process, and the more involved you are in the learning process, the more you will learn. Also, as you read your math book, recite aloud in your own words what you are learning. This is an immediate review and it will help consolidate it in your memory.

If you sit in math class and take no notes, or silently read your math book without reciting aloud or taking notes, you will remember very little after a few days. We remember only 10% of what we read and 20% of what we hear. But studies show that students remember 70% of what they explain aloud to themselves or others. And, furthermore, by involving all their senses, by both saying and doing (writing, manipulating the material, working out problems), students are able to remember an incredible 90% of the material learned.

When you go to a math tutor for help and the tutor explains the material to you, who is going to remember more, you or the tutor? The tutor, of course! If the tutor is "saying and doing," the tutor's abilities are sharpened. So if you go to a tutor for help, after the material is explained and shown to you, say "Now let me explain it to you, and work out a different problem to see if I understand."

HOMEWORK

I mentioned earlier that you should always strive to complete homework assignments the same day they are assigned to ensure that your review takes place as soon after classroom learning as possible. Make sure to read your assignment before you tackle your homework. In an effort to save time, many students attempt to do their homework problems without reading the topic. In the long run, this turns out to be a wasted effort, because they miss essential details and must start all over. Nothing wastes more time than to have to do something twice! So be methodical. Work on all the problems that were assigned to you, and even more. The more problems you work out, the more your confidence, competence, and speed in doing math will build. It may seem like a lot at first, but your increasing fluency and self-esteem will reward you handsomely.

PROBLEM SOLVING

Read each one of your homework problems carefully, at least two or three times. Do you understand it? Can you state it in your own words? Determine what things are given, what are the unknowns, what relationships exist, and what the problem asks for. Write these down. Now ascertain how to achieve the results from what you already have. Look at other problems you've learned and see if the same procedures can be used. Try to locate simpler problems that are similar to this problem. Draw out the problem. Make tables, illustrations, diagrams, and so on. This will give you direction for your analysis and computation. An illustration often can clarify the meaning of a difficult problem or formula. If you're a visual learner, it will be a double bonus for you. Write the equations needed. Estimate the answer and decide the operations that have to be done. Then do the necessary manipulations, checking yourself step by step. Once you are finished, use any of the checking procedures described next.

THE TEN COMMANDMENTS OF WORK CHECKING

Many students lose a significant number of points on their homework problems and on tests because of errors in simple computation, not because of a lack of understanding. If you want to get the highest score possible, it is important that you learn to check your work carefully. This process is similar to proofreading your essays in an English course. Get into the habit of checking all your homework problems as if you were taking a test. It will be good practice for the real thing. Use the following guidelines.

1. *Does your answer make sense?* Is it reasonable? Reread the problem to be sure you have approached it in a logical, systematic way.

2. *Does your answer fit your estimate?* When you work out problems, be sure to estimate the correct answers first. Estimate reasonable upper and lower limits; in other words, find the range within which the answer should fall. Then work the problem out and

see if the answer and the estimate are of about the same magnitude. If not, rework the problem in a different way to see if your new answer is close to the estimate.

3. *Recalculate.* Recheck your division by multiplying, your multiplication by dividing, your addition by adding the numbers in a different order, and your subtraction by adding. Many students make simple computational errors in these operations, leading to a significant point loss on exams. Others reverse or forget to carry numbers, or forget the middle term when squaring binomials; or don't do the same operation to both sides of an algebraic equation. If you use a calculator for these operations, make it a rule to calculate each operation twice. Always check and make sure you are using the right order of operations when solving algebraic equations. When factoring using exponents, are you applying exponents to all the factors? Have you multiplied the exponents when raising an exponential form to a power? When dealing with radicals have you applied the radical to every factor inside the radical?

4. *Do your problem twice.* Some students find it helpful to complete the problem and then to do it over again (using an alternative method, if possible) without looking at their previous solution. If the two answers are different, they know that one solution is definitely incorrect. At this point, they go over each step and carefully check for errors. Other students prefer to check for errors at each step along the way during problem solving.

5. *Check your usage of signs.* When you multiplied two negative numbers, did you get a positive one? Did you change a negative sign to a positive one when you switched it from one side of an equation to the other?

6. *Check your decimal points.* Did you put the decimal point in the right spot? If you estimated the correct answer, it will be easier to check if your decimal point is properly placed.

7. *Recheck your writing.* When you worked out problems on scrap paper, did you transfer them correctly to the answer sheet?

8. *Check your exponents.* Are you handling exponents correctly, multiplying, dividing, adding, or subtracting the necessary values?

9. *Reread visuals.* If you are required to read charts, tables, figures, or graphs, do you double check everything you are looking for?

10. *Substitute your answer.* Does your answer satisfy the given conditions of the problem? After working out the problem, take the answer you got and substitute it for the unknown quantity in the problem. If it doesn't fit, double check your calculations or work it out a different way.

EXERCISE 8-2: CHECKING PROCEDURES (*optional*)

Part 1

Identify at least three more checking procedures that are helpful in preventing math errors.

1.

2.

3.

Part 2

List the three most useful procedures for checking the math you are currently working on.

1.

2.

3.

PRACTICE, PRACTICE, PRACTICE

Nothing succeeds like excess. Work out lots and lots of sample problems: practice, practice, practice. Make problem solving a part of every study session. This is your most powerful aid to effective math learning. As a rule of thumb, work out at least ten problems per study session and review at least five problems from previous study sessions.

Your proficiency in solving math problems increases with practice. Cover up the solutions for problems solved in your text and work out the problems yourself. Study math every day if possible, or at least five times a week. The more you review and work out problems, the more you will learn and the more comfortable you will become.

OVERLEARN YOUR MATH

Since math constantly builds upon itself, each successive layer of new information must rest firmly on previous layers. As you assimilate more and more math, you must firmly establish the principles and concepts you are learning so they will be solid supports for the knowledge that is to come later. Overlearning is an important study skill to solidify your math building blocks.

Constantly test and retest yourself on the material you absorb. Learn to recognize your material no matter the order in which it is presented to you on a math test, no matter how difficult it may seem, and no matter how it may be "disguised." When you study, don't stop after you've just barely received the information. Apply the principles you've learned in different situations; work out mountains of problems. Get to know the concepts inside and out. In addition, it's important to review previous knowledge and topics you've learned. This is another reason that, at each study session, I encourage you to work out at least ten current math problems and five review problems. This will ensure that prior building blocks remain firmly in place and don't erode with the passage of time.

FOLLOW YOUR ALERTNESS CYCLES

Complete Exercise 7-2 from the previous chapter to determine when your energy level is at it's highest, and to plan your study sessions at this time. I call this "rolling with your alertness cycles." However, many students put their math studying off until all their other work is completed and, by this time, they are often tired. This is like running up a down escalator with your arms full of books. You get exhausted and get nowhere fast! So be sure to study math *before* all your other subjects. By studying math first, when you're most alert, you will be a more efficient learner. The concepts will come to you more readily.

TAKE BREAKS

Many studies have shown the distinct benefits of distributed practice—short study sessions interspersed with rest breaks—over massed practice—one long, continuous session without a rest. In math, several short study sessions of intense work and concentration are much more effective than sitting for hours trying to figure out problems. I recommend that you study for no longer than 20 to 40 minutes and then take a 5- to 10-minute study break. During your break, stand up, move around, stretch, get something to drink, and then return to your studying. After 2 hours of study, take a much longer break, for perhaps 20 minutes. This regimen of short study periods with small breaks prevents mental, physical, and emotional fatigue and keeps your motivation high. You'll find that, even during your short 5- to 10-minute breaks, your mind will be working on math and, when you come back to studying, your work might even go more easily for you.

Janice, a nursing student, told me she was working with a study partner for three hours trying to solve a math problem. They finally decided to give up, and she walked her friend to the door and said good-bye. When she came back to her desk and looked at the problem, the answer came to her immediately!

Ed and Nick, two brothers studying electronics, came for study-skills help. They said they had studied electronics math for five hours together and, at the end of that time, they felt they knew less than they knew at the beginning of the session. I told them I wanted them to make only one change in their study habits. I still wanted them to study for five hours, but to spread their studying into five 1-hour study periods distributed over two days, and to interrupt their 1-hour study periods with a 5- or 10-minute break every half-hour. A few weeks later they reported that this made all the difference. Now they were learning and retaining everything they studied.

 STOP PROCRASTINATING—DO IT NOW!

Are you a motivation killer? Do you like to put off doing today what can be done tomorrow? If you tend to procrastinate and put off your math studying, it's time to take *Super action.* Super action is my four-part formula to overcome procrastination. Here's how it works.

PART 1: KEEP TAKING LITTLE BITES

Many students procrastinate because their math assignments seem too large, overwhelming, or complicated. To deal with this problem I encourage you to dissect your studying into small "bite-size" pieces. This would result in a series of subtasks. For example, you might decide to read your assignment in small sections or separate it into different topics, theorems, or formulas, and then practice each separately.

Work on one subtask at a time. Arrange it so that each task takes you only a short time to accomplish. If the assignment is particularly hard, you could spread the tasks out, taking lots of breaks between tasks or perhaps completing them in more than one study session. You will accomplish quite a bit if you work on difficult assignments in small pieces or in short blocks of time. Your work will go much faster and, before you know it, the whole assignment will be completed.

Allen, a trigonometry student, found this technique really helped him to stop procrastinating on his math homework. He said he felt like a small mouse who was determined to finish a large hunk of delicious Swiss cheese. Bit-by-bit, piece-by-piece, he took lots of little bite-size pieces and, before he knew it, he got the whole job done.

PART 2: REWARD YOURSELF FOR GOOD STUDY AND CONCENTRATION

Set up a reward system for yourself. It can increase your motivation for studying math and help discourage procrastination. A principle we learn from behavior modification is that if a behavior is rewarded *after* it is performed, it is more likely to be continued.

Many students feel their reward is the grade on the test they are studying for, but this is not reinforcing enough for most students. A test may or may not turn out the way you want it to. I believe that more important than the grade on the test is the fact that you have studied and concentrated well. So stop procrastination cold! Reward yourself when you accomplish what you've set out to accomplish. Exercise 8-3 teaches you how to set up your own reward system.

EXERCISE 8-3: YOUR MATH STUDY REWARD SYSTEM

To begin your reward system, use a piece of half-inch graph paper, the kind you may have used in elementary or junior high school. If you can't find this type of paper, rule half-inch squares on a blank sheet of paper. Assign each half-inch box the value of 15 minutes. This becomes your "study bank." Then, when you study math, note the times when you begin and when you stop. For each 15 minutes of good concentration, put a diagonal slash through one of the half-inch boxes.

Next, fill out the reward menu found below. Decide on what each 15-minute block or combination of blocks is worth to you. Make a list of rewards that you find appealing and are important to you. For example, you may decide that one 15-minute block is worth an apple, or four blocks are worth a half-hour phone call to a friend. Your rewards can be as simple or as elaborate as you want. You're doing this all for yourself, so choose things that you find enjoyable and that you are willing to work for.

When you study, credit yourself with each completed 15 minutes of study time by putting a diagonal slash through one box—an hour's worth of study accumulates four boxes, and so forth. When you are ready to reward yourself, consult your reward menu to see how many boxes you have to "cash in" to get the reward you want. If you've accumulated enough, debit your study bank by placing a slash in the opposite direction (creating an *X*) through the boxes you're using up. Try out various rewards on your menu and see which are most powerful to prevent procrastination and increase motivation. The fruits of your labors will be sweet, both literally and figuratively.

My Personal Reward Menu

Decide for yourself what your studying time is worth to you. Each unit of 15 minutes or combination of these units may be used to reward yourself with something special. You decide!

Activity or Item	Unit Value
Examples:	
Snack	1
Phone conversation (30 min.)	4
Movie	8
A swim	8
____________________	______
____________________	______
____________________	______
____________________	______
____________________	______
____________________	______

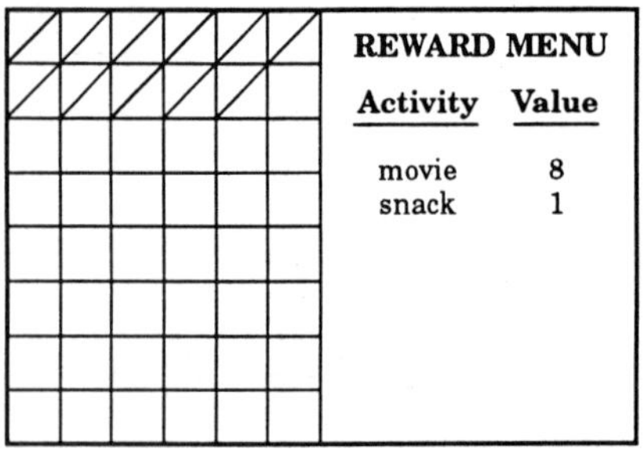

REWARD MENU

Activity	Value
movie	8
snack	1

Running total representing eleven 15-minute study intervals (blocks) credited in your account or "study bank."

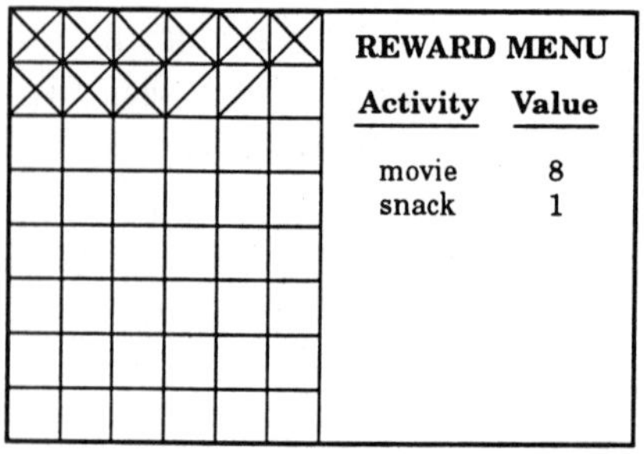

REWARD MENU

Activity	Value
movie	8
snack	1

As your reward, you've cashed in eight 15-minute study blocks for a movie and one 15-minute block for a snack. This leaves a credit of 2 15-minute blocks remaining in your "study bank."

PART 3: USE A KITCHEN TIMER

Did you ever imagine that a simple kitchen timer could stop procrastination in it's tracks? It works miracles for people who use it.

If you have trouble sitting down to study, set your kitchen timer for 15 minutes or half an hour, depending on how you're feeling. Make sure the timer is in a different room from where you are studying. Now tell yourself: "I only have to study for this period of time. When the buzzer goes off, I'll go and shut it off and, at that time, I'll decide whether I will set the timer for another small block of time." This puts studying totally in your control, and it forces you to move around and take small breaks. Even if you study for at least one 15-minute block, it's a lot better than procrastinating all day. I've also found that, once students do start to study in little blocks of time, they begin to get really involved in the material and they often decide to reset the timer, and study more. Try it—you might like it!

PART 4: CHANGE YOUR INNER SELF-TALK

Many procrastinators get stuck in negative, self-defeating self-talk. I urge you to tune into this inner dialogue so you can challenge it with super action self-talk. Here are examples of shifting this negative dialogue to action-oriented self-talk. After reading the eight examples of super action shifts in Exercise 8-4, add at least two of your own.

EXERCISE 8-4: SHIFTING TO SUPER ACTION SELF-TALK

Procrastination Self-Talk	Super Action Self-Talk
1. I don't know where to start.	I'll divide my assignment into small chunks and work on only one chunk at a time.
2. I don't know how to do it.	I'll look up the information in other math books and see how they explain it. I'm bound to get it!

	Procrastination Self-Talk *(continued)*	**Super Action Self-Talk** *(continued)*
3.	I'll wait until I can ask the teacher in class.	I'll work on some of it now and then I'll have better questions to ask.
4.	I'm just not in the mood to do it.	Do it now! Just do it!
5.	Oh, I still have time; I can wait until later.	Time has a way of running out quickly; get it over with *now*!
6.	There's just too much to do.	Take one step at a time; Rome wasn't built in a day.
7.	I feel bad, but I keep wanting put this off.	If I do even a little bit of work on this, I know I'll feel better.
8.	I prefer doing my favorite subject first. I can always do math later.	Once I get into my math, it'll begin to be more enjoyable and fun.
9.	(add your own)	(add your own)
10.	(add your own)	(add your own)

So make up your mind to get things done *now*! It's a mental blueprint for success, not just in math class but in every job or career. Employers love what super action accomplishes. It's reinforcing, and you'll get hooked on it, too.

TACKLE YOUR MATH BOOK

Paula, a young woman taking a developmental math course, came to me because she was struggling in her math course. As we talked, Paula confided to me that she had never read her math book. She said she depended on the teacher's lectures to give her all the information she needed for the exam. That was a major mistake. By not reading her math book, she missed crucial information that supported what was being taught in class. The more ways you can be exposed to math—the more you read, the more problems you work out, the more you do—the more you'll learn!

READ THE CHAPTER BEFORE AND AFTER CLASS

Make it a practice to read over the assigned chapter *before* you attend math class. Many students avoid reading their math book or wait until after they've heard their instructor's explanation before they read their assignment. If you are one of these students, I encourage you to take heed: read your chapter prior to class even if you have little understanding of it. Look over the topics, diagrams, charts, terminology, formulas, and examples. This familiarizes you with the topics your teacher will present in class, and it prepares you to understand and absorb the material presented in class more readily. Pose questions on any confusing or difficult ideas in the chapter that can be answered in class, and then listen carefully to your teacher's explanation. You'll find that you become a better listener and you'll learn much more from the class than previously. After attending class, first read your math assignment in depth, and then do your homework problems. This will further your understanding of the math topics discussed in class and will also be an excellent review.

HOW TO READ YOUR MATH TEXT

Math reading assignments need to be tackled at least three times. Before you panic, let me explain; it's amazingly simple and straightforward. First, survey your assigned material. When you survey, read the lead paragraphs, the first sentence of each paragraph throughout the section, and the closing or summary paragraphs. Read all highlighted areas, tips, subtitles, illustrations, charts, and graphs. This will give you a basic idea of what the section is about. As you do this, pose questions that you believe your text material attempts to answer or that you believe might be answered in your math class. Be sure to complete this survey before you go to class.

Your second reading is your in-depth one, where you'll read for mastery, reread sections for understanding, and mark and underline in your book. Have you noticed that reading your math book is quite different from reading texts in psychology, humanities, English, or anthropology? In these other areas, as much as 80% of the actual words are unimportant, except for the fact that they are needed to link relevant ideas together. When you read math, *every word is important.* Math books are usually written succinctly and to the point. Each word is carefully chosen to explain a concept. You need to read a math text slowly, carefully, and with good concentration. Don't rush. Work on grasping each concept before going on. Learning math takes time. Reread the concepts, several times if necessary, until you gain mastery. If, after several attempts, you still can't understand a particular sentence, review the topic up to that point and then reread that sentence. You also can formulate questions on these difficult points that you later can ask your instructor or a math tutor, or look up in other math texts.

In other subjects, you may be able jump around in your text, skip chapters, or read only parts of chapters, and still understand the material. In math, each idea builds upon another; each topic presented assumes you've mastered previous topics. You can't jump into the middle of a chapter or skip a few chapters and expect things to make sense. So move progressively in your math text from front to back, mastering topic by topic.

A picture says a thousand words—as do diagrams, charts, illustrations, and figures. *Don't skip over them.* Work to comprehend what they are illustrating.

Write and recite as you read. Some students hesitate to write in their math books which, they fear, would then lose resale value. But many bookstores don't care if the books contain handwritten notes or underlining, and you could always do this in pencil. Let me assure you, the best math students always mark in their texts! Writing as you read gets you more involved in the reading process. Remember what I noted earlier: we retain only 10% of what we read but 90% of what we say and do. Make notes to yourself directly in your text. Whenever you encounter new symbols or terms, look up their meaning and write them down in the margin. It's important that you know your math terminology. Label or mark in the margin important ideas, formulas, procedures. Underline main points. Write out questions you have about the material. Cover up important topics and then explain them aloud in your own words to yourself or to someone else. Keep testing and retesting your understanding and retention of each topic as you read. View your text as one of the main sources for the types of questions that you are likely to have on your math exam.

Examples are an integral part of each math assignment. Go over each step methodically until you understand the terms used, and be sure to follow the reasoning used for each procedure. Afterward, cover up the solutions and visualize, say aloud, or carefully work each problem out step by step, by yourself, to see if you conceptualize the procedures. When you finish, compare your steps to the sample steps in the text. Design other little tasks for yourself as you read. Work out all proofs and derivations. Work out as many problems as you can to ensure that you assimilate the topics being presented. You will find it easier to learn facts when you comprehend them, rather than follow them by rote.

The third reading of your chapter is for review and taking notes. Write down all important ideas, facts, equations, theorems, examples, and summaries in your notebook or on index cards. Work out your homework problems and do lots of practice problems.

Don't feel overwhelmed! When you learn to drive, there's a lot to attend to and some bumps along the way. But practice brings familiarity, then a sure touch, next comfort with increased speed and precise handling, and then smooth sailing. The same will happen with math.

MEMORIZE TERMS, RULES, PRINCIPLES, FORMULAS

To memorize a rule or principle, first read it and make sure you completely understand every part of it. Then write a sample equation to show how it's used, or illustrate what it represents with a picture. Once you've done this, read the definition aloud about five times, in your own words if possible. Then write the definition on a 3×5" card or in your study notes so you can review it often.

Use 3×5" flashcards to study math's own unique vocabulary. You also can use flashcards for new facts, formulas, properties, proofs, or procedures that you wish to recall quickly. Make sure you understand and know how to use all of these before you commit them to memory. If you understand the concepts behind the formulas and various procedures, you will be able to recreate them easily if your recall falters in any way. *A word to the wise:* Some time you may have to commit to memory an extremely difficult concept that you do not understand. Don't be discouraged. This even happens with students in advanced math courses. Your understanding may come later by asking questions of your instructor or tutor, looking in other texts, and, in some cases, only by taking more advanced math courses.

USE TWO OR MORE MATH BOOKS

Use two or more math texts when studying. They provide you with many more worked-out problems to study as well as numerous sample problems to work out. Make sure you have the answers listed in the back of the book or in a solutions manual. In addition, by having more math books to study from, you have different explanations for the concepts you're learning. Remember, your instructor has probably studied or taught out of several books and will be using this knowledge to teach the course and test you. You also might choose books that appeal to your perceptual learning channels. For example, if you're a visual learner, you might choose texts that have many colored diagrams, illustrations, or charts. If you're a verbal learner, you might prefer a book with more explanations, and if you're a kinesthetic/tactile learner, you might search for texts with real-life examples you can experiment with.

Elena, a college algebra student, said that when she studied, she spread out four algebra books around her, and it felt as if she had four different teachers explaining algebra to her.

By having several different books, you can get an understanding of the topics you are studying from many different viewpoints. And, if your text doesn't quite make sense to you, another might explain the topic more understandably. In addition, by working with different textbooks, you review a topic each time you read about it. To obtain a large collection of sample problems and answers for your study sessions, I suggest you look for the college course outline books prepared by Schaum, AMSCO, Barron's, or Barnes and Noble.

NOW IT'S TIME FOR YOU TO DECIDE

EXERCISE 8-5: MY COMMITMENT FOR IMPROVEMENT

List here the five most important study skills you've learned in this chapter that you're willing to incorporate into your study routine within the next month.

1.
2.
3.
4.
5.

SUMMARY

Use appropriate math study skills and you'll achieve major breakthroughs on your road to success. This section of our road map highlighted strategies for obtaining greater mastery and retention of mathematical concepts. I encouraged you to jump right in and get actively involved in the study process; to carefully select your math class and your teacher; to go to class regularly; to take complete class notes; to stay current and not fall behind; to take the initiative to ask questions; to review immediately after learning and again eight hours later; to tackle your math reading assignments in three different ways; to work out lots and lots of practice problems every day; and to take super action!

CHAPTER

9

CONQUERING TEST ANXIETY

Test taking creates enormous anxiety for many math students. Fear and overwhelming doom often come over them like an impending, life-threatening storm. Darkness fills the horizon, and they shudder in their footsteps, paralyzed, unable to take any action. Thoughts race from one thing to another and suddenly, blankness, emptiness, all that was in their heads—gone! An empty slate with no numbers, no formulas, nothing! Foreboding feelings of agony or sudden death loom heavily in the air. Dreams and ambitions melt into oblivion. Feelings of worthlessness, uselessness, and hopelessness abound. Has this ever happened to you? If so, you are not alone. Thousands of math-anxious students experience the devastating effects of test panic each day.

The symptoms of test anxiety may be as pervasive as those just described or as minimal as forgetting an easy formula you've used hundreds of times. Either way, your ability to perform well on math exams becomes terribly handicapped. In some cases, you may become totally immobilized.

EXERCISE 9-1: SYMPTOMS OF TEST ANXIETY

The following checklist will alert you to the physical and mental signs of test anxiety. Check off any of the symptoms, mild or severe, you have experienced before or during tests.

PHYSICAL		MENTAL	
increased sweating	________	confusion, disorganization	____
increased need to urinate	_____	foggy thinking	__________
headaches	__________	blank mind, freezing up	______
shakiness	__________	overwhelming fear or panic	____
upset stomach	________	poor attention span	________
pounding heart	________	poor concentration	________
loss of appetite	________	increased errors	________
tightness of muscles	______	fleeting thought processes	____
stiff neck	__________	narrowed perceptions	______
backaches	__________	immobilized creativity	______
fatigue	__________	nervous worrying	________
"free floating" anxiety	______	pervasive negativism	______
insomnia	__________	weakened logical thinking	_____
total mental fatigue	________	feelings of impending doom	___
feelings of inadequacy	______	distracting thoughts	________

Kirk, a college freshman, came to me and pleaded: "Why me? I studied and studied. I knew my information, but I just blew it. I feel so dumb. I confused formulas, multiplied 2×4 and got 9, and on top of it all, copied my answers incorrectly to the answer sheet." Why, indeed?

The roots of test anxiety are threefold, and any or all may be the culprit. They include poor test preparation and test-taking strategies, psychological pressures, and poor health habits. In this section of our road map, we will explore how to deal with each of these trouble-makers.

TEST PREPARATION AND TEST-TAKING STRATEGIES

Many students feel anxiety about taking math exams because they have poor study skills and inadequate test-taking strategies. The previous chapter discussed effective math study skills. Good skills are essential for comprehensive and long-term recall in math. In addition, some specific test preparation and test-taking strategies make a world of difference. They relieve much of the pretest jitters and insecurities that anxious students experience.

Here are some important strategies I recommend for students.

1. When a test is announced, check in advance its format and content. Know what your instructor expects. Make a detailed list of all the topics the test will cover. Be sure to find out how many questions it will include and the time allowed. Will you need a calculator, charts, tables, or scrap paper? For example, Michelle learned that her statistics final was going to include 60 questions and that she would have 3 hours to complete the exam. Each question would have equal point value, so she'd have three minutes to answer each question. She learned that even if some of her answers were wrong, she could get partial credit if she showed her work. She found out what six major topics the exam would cover. She'd be allowed to use pencil and bring lots of scrap paper and her calculator. Statistical tables would be provided for all students.

EXERCISE 9-2: YOUR NEXT MATH EXAM

Can you answer each of the following questions about your next math test?

1. How many questions will be on the test?
2. How much time will be allowed?
3. Will each question have the same point value?
4. What topics will be covered in the exam?
5. Can you bring a calculator, scrap paper, or other aids?

2. Review each concept that the exam will cover. Locate problems that illustrate these concepts. Learn how to recognize these problems when they're placed in random order, are worded differently, or the concepts appear disguised. Solve lots of difficult problems on each

topic tested. Work on gaining mastery of each math concept you're studying. I encourage you to know your information inside and out, so you can practically do it in your sleep. Then, no matter how the information is presented to you on the exam, it won't throw you, and you'll recognize it no matter how it's camouflaged.

3. After completing your review, design your own sample practice exams covering each topic you've mastered. Choose questions from your text, workbook, other math texts, tests from past semesters, study guides, or college course outlines series (for example, Barron's, Barnes and Nobles, AMSCO or Schaum). Make sure that you have available to you the correct answers for these questions. Plan practice exams that include the same number and type of questions as your test. Design your practice exams carefully so they accurately reflect what has been done in class, what your teacher expects, and areas in which you may need more work.

4. Set a time limit for your practice exam, giving yourself the same amount of time allowed for on the test. Then take several timed practice exams. Time yourself with a kitchen timer or an alarm. Many students can figure out all sorts of complicated problems if given enough time, but then they panic under the time pressures of the test. So be sure to set up your own timed exams. And practice, practice, practice. Work on building up speed and accuracy. If possible, take these practice exams in a room similar to the testing room.

EXERCISE 9-3: DESIGN PRACTICE MATH EXAMS

1. List the sources you will use to get sample problems.

2. How will you get the correct answers to each problem you select?
3. How many questions are you going to include in your practice exams?
4. How much time will you allow yourself to complete each question?
5. Where will you take your practice exams?

5. Use your practice exams to begin to recognize problems out of the context of the textbook or your class notes. Note the types of questions that cause you difficulty. Give yourself more practice in these areas of difficulty. Ask your teacher for further help or go to tutorial or help sessions.

6. Use your practice exams to analyze what typical errors you are making. Identify whether there's any pattern to the errors. For example, do you consistently use one of your formulas incorrectly, or do you often forget to carry numbers during calculations? Perhaps you often put the decimal in the wrong place, or use signs incorrectly, or reverse numbers when recopying them. Perhaps you often multiply instead of adding exponents when multiplying exponential forms with like bases; or maybe you confuse the order of operations when solving algebraic problems. Make a list of these errors. Your list will be very important later when proofreading your math test.

EXERCISE 9-4: YOUR TYPICAL MATH ERRORS

Do an error analysis. Look over your past math exams and your practice exams and list the typical careless errors you have made:

7. Get plenty of sleep in the 48 hours prior to your exam, and eat healthy meals (see the next section on preperformance health tips). Make sure to practice deep breathing techniques in the half-hour before the exam. Arrive at the exam just on time and discuss the exam with no one.

8. As soon as you receive your math exam, write on the top corner of the exam paper *all* the formulas, rules, and key information you'll need. Do this first, before beginning your math exam. This way, you'll have a handy reference guide—your "cheat sheet"—in front of you, and you won't have to keep it stored in your head. Most students find this to be a very helpful technique that allows the exam to go more smoothly for them.

EXERCISE 9-5: YOUR MATH "CHEAT SHEET" (*optional*)

List here all the formulas, rules, principles, or key ideas you will need to remember for your next math exam.

9. Note the time limit and the point values for each question, particularly those that carry more value. Adjust your time accordingly, allowing more time for questions having higher point values.
10. Circle or underline significant words in each problem. Read each problem slowly and carefully so you don't misinterpret anything.
11. If you are very anxious, you may find it helpful to start the math test with the easiest problems first. These may not necessarily be the first ones on the test. Do whichever ones come easily for you. You need not go in order. Then go to the next easiest, then the next easiest, working your way on up to the more difficult ones as your confidence level increases.
12. Don't spend too much time on any one problem. Skip the harder ones, but mark them so you can find them readily. Return to these difficult questions later as your confidence builds. Also, be careful not to get stuck for a long time trying to remember a step on any one problem, even if it seems easy. Don't allow yourself to lose too much valuable time on any individual problem that would prevent you from completing the exam. Go on and the key step you are trying to remember will probably come to you later.
13. Throughout the exam, focus on remaining calm, relaxed, and positive. Check your breathing often; keep it regular and slow. Check your neck and shoulder muscles and loosen tight areas. Say positive statements to yourself and push away any disturbing or distracting thoughts.
14. If you can use calculators, take advantage of them. Learn how to use them well ahead of time. There are appropriate and inappropriate times to use them. They can be very useful for checking your answers.

15. When working on a problem, write down in an organized fashion whatever you know. This might help you to figure the problem out, and some teachers give partial credit in recognition of what you do know. Don't skip a question if you know even a little about it. The answer might come to you as you work on it. But save these more difficult questions until the end of your test.

16. Allow yourself the *entire* test period to finish the exam. Don't allow yourself to get nervous when some students leave the exam early. Teachers have found that students who leave the earliest are often the ones who do the most poorly on exams.

17. After finishing the test, use the remaining time to verify your solutions. First, you should check for reasonableness. Ask yourself: Does the number fit? Does the answer make sense? For example, can Marvin's little sister really weigh 658 pounds? Or did Martha realistically spend $12,490.00 on a week's worth of groceries? Second, check the specifics of the problem. Have you satisfied all the conditions? If you are unsure of your answer, do it again. Use the ten commandments of work checking that were given to you in Chapter 8.

18. Lastly, proofread your test. Did you answer all the questions? Are there any parts you missed? Look for any omissions. Check to see if you've made any of the typical errors or pattern of errors you identified previously (see Exercise 9-4).

19. Leave, and reward yourself for a job well done!

DEALING WITH PSYCHOLOGICAL PRESSURES AND FEAR

Psychological factors contribute dramatically to test anxiety. Students often talk negatively to themselves, put themselves down, and experience low self-esteem. No wonder they may confront difficult, high-stress situations poorly. Whether realistic or not, some students create high expectations for themselves and fear that failure or a poor grade will jeopardize their entire future. Family members or loved ones, often unintentionally, may add to this burden with excessive demands to do well. Encumbered by enormous self-imposed and externally imposed pressures to succeed, students are likely to panic at the thought of what might happen if they don't.

Michael was one such student. He gave up his well-paying electronics job to return to school and finish the engineering degree he had started 12 years earlier. His wife became the sole breadwinner, supporting him and their three children. Everyone, including Michael's parents, was excited for him and expected him to be a great success. But after three semesters in school, Michael was very discouraged. He was repeating one of his math courses for the third time. It was at this time he came to me for counseling. He had just panicked on an important exam and failed it. Michael appeared to be an intelligent, well-prepared, mature student, but his anxiety level had skyrocketed. He felt like a failure. He was sure he was letting down his family.

Research has shown that performance is a very sensitive barometer to the level of anxiety students experience, but the relationship between performance and anxiety is not linear. You also may be surprised to learn that the effects of anxiety are not uniformly negative. In fact, research has repeatedly shown that some anxiety is necessary for optimal performance. When plotted in the form of a graph, the anxiety/performance curve takes the shape of an upside-down U (see Figure 9-1). If students feel low anxiety, they are inadequately motivated, accomplish little, and perform poorly. As anxiety increases to a moderate level, perceptions sharpen. Students feel alert, energetic, clear-minded, motivated, and at their creative best. Not surprisingly, performance reaches an optimal level. But if anxiety continues to rise to higher and higher levels, the effect is catastrophic. Students become increasingly indecisive, make careless errors, and show poor judgment. Overwhelmed by anxiety, students become disorganized, and they can neither function effectively nor think clearly. Performance plummets until, at the highest levels of anxiety, they freeze up, their minds go blank, and total panic sets in.

Once begun, this process is self-perpetuating and extremely difficult, if not impossible, to reverse. Students caught in this bind may have no choice but to abandon the situation. One student I worked with said that he totally blanked out on his exam and sat for 30 minutes in a state of panic. He finally left and decided to go home. As

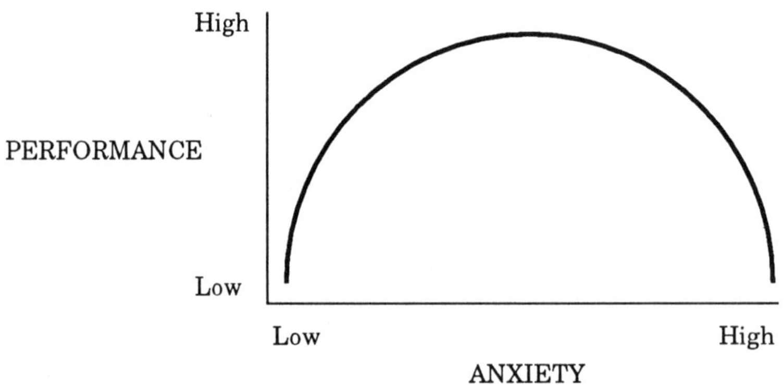

FIGURE 9-1 ANXIETY PERFORMANCE GRAPH

he opened the door of his car, all the information he needed for the test came flowing into his mind. He could see all the formulas, all the answers to the problems he had just been staring at. It was as if the doors of memory opened, the flood gates burst and everything came rushing in. But it was too late!

The key to sustained successful performance on exams is keeping your anxiety within a manageable level, thus preventing it from ever reaching destructive proportions. You may do this by using a three-pronged approach: (1) reversing negative self-talk, (2) using relaxing and calming techniques, and (3) through visualization.

REVERSING NEGATIVE SELF-TALK

The first target in test anxiety reduction is the most devastating of all psychological culprits: negative self-talk. I have found that most students who suffer test panic repeat a continuous stream of negative, self-defeating statements to themselves before, and especially during, an exam. Here are some examples of statements students often say:

I can't do it.
I'll never finish all this in time.
It's too difficult for me.
I feel so dumb when it comes to math.
Everyone else knows how to do this but me.

I just don't have the ability.
Who am I fooling? I don't know what I'm doing.
I've had trouble with math all my life.
I hate tests.
I hate this course; I wouldn't be here if it weren't required.
Why do I need to learn this anyway?
I'm too nervous to concentrate.
Maybe I should just drop this course.
The teacher's sure to ask me things I don't know, and I'll blow it again.
Who cares about this test (course) anyway?
I always do well on the homework, but the test is never like the homework.
The problem looks too easy; I probably got it wrong.

Statements like these tend to become self-fulfilling prophecies. Students begin to focus their attention and energy on this continuous stream of inner dialogue. They start to think less clearly, concentration diminishes, attention wanders, and they have less energy for dealing with difficult questions on the exam. More and more, they become distracted by disturbing or intrusive thoughts. Daydreams about being in a more pleasant environment—any other place than where they are—are common. Slight noises or movements in the exam room soon become monumental distractions. Anxiety levels shoot way out of control.

How can you deal with these negative self-statements? The most crucial step is to become aware of the negative mind talk and its ability to sabotage your efforts for success on an exam. Most students are not even aware of their negative inner dialogue, let alone the devastating effects it has on their performance.

EXERCISE 9-6: BECOMING AWARE OF CONTINUOUS INNER DIALOGUE

Tune in to your thoughts and observe the little voices in your mind. What do you hear in your mind when you worry or fret about an upcoming exam? What thoughts get you nervous? What are your worst fears? What thoughts seem to stop you in your tracks and sabotage you? What negative things are you telling yourself? Write them down here.

Once you identify your negative self-statements and bring them into the open, you can deal with them more rationally and effectively. The next step is to challenge each of these negative, self-defeating statements and reverse them to positive, self-enhancing statements. Examples of negative to positive shifts include the following:

Negative Self-Statements	**Positive Reversals**
I can't do it.	I know I can do it.
I feel overwhelmed.	I can do this one step at a time.
It's too difficult for me.	I have the aptitude to learn this.
I'm stupid.	I have good abilities.
Everyone else knows how to do this but me.	I am learning how to do this.
I'm not really smart.	I know I am capable.
I'm too nervous to concentrate.	I am remaining calm and relaxed even under pressure.
I'll flunk, so why try?	I am learning more each day; success is bound to follow.
I hate tests.	Tests are becoming easier for me.
Who cares about this test anyway?	This test is a positive challenge for me to show what I've learned.

EXERCISE 9-7: NEGATIVE TO POSITIVE SELF-STATEMENT SHIFTS

In the following spaces, create your own negative to positive shifts. List the negative thoughts that make you anxious before an exam, and then shift them to positive, achievement-enhancing ones. Be sure to use only positive terms. Avoid all conditional statements ("should" and "could"), and write the positive statements in the present tense. Although you may not at the moment believe the positive statement is true for you, it is nevertheless necessary to state it: no "wills", "ifs," or future tense.

Negative Self-Statements	**Positive Self-Statements**
______________________	______________________
______________________	______________________
______________________	______________________
______________________	______________________
______________________	______________________
______________________	______________________

This process of changing negative to positive self-statements will help to demystify and disempower your negative self-talk, and eventually it will fall by the wayside. Every time a negative thought or self-statement reappears, you should reverse it to a positive statement.

If, despite all efforts, some negative thoughts continue to plague you, a useful technique to overcome them is the thought-stoppage procedure.

EXERCISE 9-8: THOUGHT STOPPAGE TECHNIQUE

Use this technique whenever a disturbing or distracting thought comes to mind. First, in your mind, you are to yell "STOP" and then say "calm." Then deliberately relax by taking several deep, comfortable breaths. The aim is to get at least a momentary break in the negative thought process, making it possible to achieve conscious control and regulation of self-defeating inner dialogue. Each time the negative or disturbing thought comes to mind, repeat this procedure: yelling "STOP," saying "calm," and then deliberately relaxing. Allow this technique to become automatic, and soon the negative thoughts will be completely eliminated.

EXERCISE 9-9: POSITIVE STATEMENTS FOR ACADEMIC SUCCESS

To keep inner dialogue positive and reinforcing, repeatedly conjure up positive statements and visualize them being true. In the list below, check off the positive statements you think would be most rewarding to you (check at least five). Be sure to add some of your own.

_____ I'm a good student.

_____ I'm learning more each day.

_____ I am capable.

_____ I have good abilities.

_____ I allow myself to relax while studying.

_____ My memory is improving each day.

_____ My mind is clear and alert.

_____ I see myself accomplishing my goals.

_____ I eliminate all distracting and disturbing thoughts.

_____ My comprehension is excellent.

_____ I am an intelligent, talented person.

_____ I have confidence in myself.

_____ I have good concentration.

_____ I'm remaining calm and relaxed even under stress.

_____ I'm handling my studies well.

_____ I can manage my workload.

_____ I'm getting better each day.

_____ I have studied well, and I know my material.

_____ I'm motivated.

_____ I can do it!

_____ I'm making it!

_____ I remain calm, relaxed, and alert on my math exam.

_____ I remain clear, capable, and confident on my math exam.

My positive statements:

1. ____

2. ____

3. ____

4. ____

5. ____

6. ____

I ask students to write positive statements such as the ones in Exercise 9-9 and post them on their mirror, refrigerator, or in the place they usually study for exams. Soon magical things begin to happen. Students report that they feel better, they study better, and they're calmer during tests. Say and visualize positive statements and you, too, can create self-fulfilling prophecies. You will begin to feel better. Your attitude will improve, and your anxiety level will decrease. You will be able to function better and more efficiently. Always keep in mind that whenever negative self-statements creep into your thoughts, either during studying or test taking, you must push them away, tell them to stop, and then deliberately relax. You must focus on relaxing, remaining positive and refocusing your attention on your studies or the test at hand.

RELAXATION

This brings us to the second part of my three-pronged approach for dealing with the psychological aspects of test anxiety: relaxation. One of the most important things you can do in the weeks, days, and even minutes before an exam is to practice calming and relaxation techniques. You should *not* study up to the last minute before an exam. Repeatedly, I've seen students who did this and then went into the exam and panicked or completely blanked out. Instead, in the last half-hour before the exam, I recommend that you calm yourself down using techniques such as the calming breath or deep abdominal breathing described in Chapter 3. By practicing slow breathing from deep within your lower lungs, you begin to gain control over your anxiety response, preventing it from accelerating and escalating to exceedingly high levels.

EXERCISE 9-10: PRACTICE THE CALMING TECHNIQUES (See Chapter 3)

Practice a deep, slow breathing pattern for a few minutes each day in the days before your exam and in the half-hour prior to your exam. Fill your lungs completely, feel your abdomen expand, hold the air a moment, and then slowly exhale.

Test-anxious students find it especially beneficial to keep to themselves before the exam. In other words, don't get to the exam too early, and avoid talking with classmates, if possible. This is important for two reasons. First, it prevents you from picking up other students' anxiety or from becoming anxious because of something they might say. Secondly, the first recall after studying is usually the best. I've met students who reported that someone asked them a question before the exam, and they responded correctly, but then proceeded to get the problem wrong on the exam. So be a hermit! When you do arrive for the exam, find a comfortable seat and begin to focus on being calm, positive, and relaxed.

Maintain a calm, slow breathing pattern throughout the exam, while continuing to say positive self-statements. Push away all distracting or disturbing thoughts and focus on staying positive and calm. If you find that anxiety begins to intrude at any time during the exam, stop and do deep abdominal breathing.

"OK CLASS - READY FOR YOUR TESTS?"

VISUALIZATION

The final part of my approach for dealing with the psychological aspects of test anxiety uses visualization. Here, while in a deep state of relaxation, you are to visualize yourself feeling confident, capable, relaxed, and successful while taking your math exam. It is important to practice the Math Test Anxiety Reduction Visualization (Exercise 9-11) each day for at least five days before the exam. It is particularly important to practice it the evening before and the morning of the exam. You should practice this excellent technique in a peaceful, quiet, darkened place where you can be alone and undisturbed. You may wish to read this exercise onto an audiocassette tape in a slow, unhurried, and calm manner. (Be sure to pause for a few seconds each time you come to a set of three dots.) This exercise should take approximately 20 minutes. You'll then have it available for repeated use any time you need it.

EXERCISE 9-11: A MATH TEST ANXIETY REDUCTION VISUALIZATION

Begin by assuming a comfortable, relaxed position. Let your body settle down inside and, gently and slowly, close your eyes. It is easier to concentrate on relaxation with your eyes closed.

Let's begin by focusing on your breathing for a few moments. Take a deep and comfortable breath, filling your lungs completely . . . hold it a moment . . . and then . . . very, very slowly, let it out . . . slowly . . . feeling a wave of relaxation going from the top of your head all the way down to your toes. . . . Now I would like you to take another deep and comfortable breath, filling your lungs completely . . . hold it a moment . . . and then . . . once again let it out . . . very, very slowly, feeling another wave of relaxation going from the top of your head all the way down to your toes . . . slowly . . . feeling more and more relaxed.

Continue to breathe slowly, deeply, regularly. As you breathe in, concentrate first on filling the lower part of your lungs, then the middle part, and then the upper part. Feel your abdomen slowly expanding . . . now exhale very, very slowly . . . slowly, saying "relax" as you do so. Continue breathing slowly and regularly, filling your lungs completely and then so, so slowly exhaling, saying "relax" and *feeling* relaxed. Allow each breath to take you into a deeper and deeper state of relaxation. Continue alone for a few minutes. (Pause two minutes)

Feel your mind and body becoming more and more peaceful and serene. . . . And now, beginning at the top of your head, allow all the tension to leave your body, relaxing further and further. Let go of all the tension in your scalp, forehead and the tiny muscles around your eyes. . . . Relax your jaws. . . . Allow your neck and shoulder muscles to go loose and limp. . . . Continue to breathe slowly and deeply, saying "relax" with every exhalation. . . . Feel all the tension drain out of your arms and hands. . . . Now allow your chest, abdomen, and back to relax. . . . Feel the relaxation flowing through and soothing each and every vertebra of your spinal column, from the top all the way down to the bottom. (Pause 30 seconds.) Now feel the tension draining from your hips, down your legs, and out the bottoms of your feet. You are relaxing more and more with every breath you take. (Pause 20 seconds.)

Now say to yourself and visualize the following (pause five to eight seconds between statements):

I see myself accomplishing my goals.
I can eliminate all distracting and disturbing thoughts.
I have confidence in myself.
I am an intelligent, talented person.

My memory is improving each day.
I see myself relaxed and calm on my math exam.
I see myself relaxed, calm, alert, and confident on my math exam.
I am relaxed, calm, capable, and confident on my exam.
I see my math exam going well for me.
My mind is calm, clear, and alert during my math exam.
I see myself succeeding in math.
I now say to myself, over and over, my own special positive affirmations and visualize them being true for me.
(Pause one minute.)

Next I would like you to imagine that you have just finished all your studying and reviewing for a math exam tomorrow. . . . You are feeling good. You have studied well, and you are confident that you will do well on the exam tomorrow. . . . You get ready for bed and you snuggle down under the blankets, but before going off to sleep you practice relaxation. You slowly and deeply breathe in and out . . . saying "relax" with every exhalation and visualizing the exam going well tomorrow. . . . You soon fall off to sleep knowing you will have a good night's sleep. . . .

Now, imagine yourself awakening the next morning. You are feeling good. You have had a good night's sleep, and you feel refreshed, clear, and alert. You eat a nutritious breakfast of low-fat protein to help sustain your alertness. You're feeling so good. You recall all the studying and reviewing you have done in the past weeks for today's exam and you *know* that you know the material. You have organized yourself well, reviewed often, and you have confidence that you will do well on the exam today. . . . Again you spend a few minutes relaxing and visualizing the exam going well today. . . .

Now imagine that you are on your way to school. What are your surroundings like? Are there clouds in the sky? What color is the sky? How does the air feel against your cheek? Now see yourself walking toward the building where your test will be given. Notice the colors, the sounds, the smells, the shadows. You are now entering the building and slowly approaching the room. You are not too early, nor too late for the exam. You have scheduled your time well in order to arrive just in time. You are like a hermit before the exam; you avoid any contacts or discussion about the test with anyone. You keep yourself in a peaceful, calm state. You slowly enter the room. You find a seat.

As you sit quietly in your chair waiting for the exam to be passed out, you close your eyes or focus them on a point of light in the distance. You tune out the other people in the room and turn inward for a few moments. You breathe slowly and deeply from your lower lungs, and you take yourself through the relaxation process you have practiced the last few days, saying "relax" with every exhalation. As you do this, you feel all the tension leave your body, dissipating into nothingness. A sense of well-being comes over you. As you sit there, you recall the studying and reviewing you did for this exam; you see your lecture notes, your textbook notes, the many problems you've worked out. You see the formulas, the important concepts, and the solutions to difficult problems. You recall all the time and effort you put into reviewing all the information, and you know that you know your work. You are feeling capable and confident. You are calm and relaxed, yet clear and alert. (Pause 15 seconds.)

Now imagine that you are given the exam. You read the directions carefully, underlining all significant words so that you won't overlook them or misinterpret them. You skim through the whole test, noticing the ideas covered, the types of problems, and the point values. You smile to yourself because you recognize all that is being asked of you, and you feel good about your preparation. (Pause 10 seconds.)

You proceed confidently through the exam, beginning with the questions that appeal to you first. You don't necessarily begin with the first question if you don't want to. You may choose the ones that seem easiest or most fun. By doing these first your confidence grows. You feel good. You feel alert, assured, and calm. (Pause 20 seconds.) You continue through the exam, pacing yourself well. You recall your notes, handouts, homework assignments, and class discussion. You proceed confidently through the exam. When you come to difficult questions that you are not sure how to answer, you leave them for a while and simply go to easier ones. As your confidence builds, you soon return to these difficult questions and the answers come to you much more readily. With some of the difficult questions, you simply sit back, take a deep breath, exhale slowly, and say to yourself, "relax." You soon find that the answers to these questions come to you much more readily and quickly. You have studied all the information. You know you know the work. You are very capable. By staying calm and relaxed, you are able to function at your highest level on the exam. You are feeling calm, relaxed, confident, and capable. (Pause 15 seconds.)

Now I would like you to imagine that you have completed the exam. You spend a few minutes checking over it for errors, omissions, and reasonableness of the solutions. (Pause 15 seconds.) You are feeling relaxed and calm and in no hurry to leave the room until you have finished checking your paper. . . . Now imagine that you hand in the exam paper. You leave the room. Once in the fresh air, you take a deep breath and exhale slowly. . . . You feel refreshed and happy. The exam has gone well and you have done the best job you could. You are feeling wonderful. You have remained calm, alert, confident, and capable throughout the exam. The exam has gone well for you, and you know that you will remain calm and relaxed for exams in the future. . . . Be sure to give yourself a special treat later in the day for handling the exam so well.

Now I would like you to slowly and gently bring yourself back to this room, carrying with you the pleasant feelings and thoughts that you experienced during this exercise.

PREPERFORMANCE HEALTH TIPS

It has been said that the body is the temple of the soul. I have found that test-anxious students often are poor overseers of the golden temples in which they reside. They eat poorly, sleep minimally, and exercise rarely. Battling fatigue and nervous exhaustion with little or no fuel or resources, they fall easy prey to their nemesis—test anxiety.

Many students get far less than the recommended seven to eight hours of sleep each night. Fatigue and exhaustion reduce efficiency and cause poor memory and recall. Time for relaxation and meditation also may be missing from a student's busy schedule. Twenty minutes of reverie or silent contemplation can often make up for a shortened night's sleep.

Students also have a tendency to stop exercising in the days or weeks before a test. A regular program of moderate exercise (for example: walking or biking 20 minutes a day) can greatly reduce stress, as well as increase alertness, clear thinking, and energy level.

For many students, caffeinated beverages such as coffee or soda are the culprit. In an attempt to be more alert for a test, students may decide to drink more caffeine than they usually do or drink caffeine when they are not normally caffeine drinkers. Unfortunately, excessive caffeine or caffeine taken by people who are not accustomed to it can only make them jittery, "hyper," or shaky for the test. If they were already a bit anxious, this added shakiness may put them over the line into panic!

Poor nutrition, such as an unbalanced diet or eating high-caloric, fatty meals, can negatively affect a student's thinking and problem-solving abilities.

Judith Wurtman's *Managing Your Mind and Mood Through Food* contains an excellent discussion on the effects that the foods we eat have on our brain's ability to process information. Tyrosine, for example, an amino acid found in protein, is the principal ingredient in two neurotransmitters. These neurotransmitters, dopamine and norepinephrine, produce alertness, clarity of mind, motivation, and drive. Research shows that with an adequate supply of these two neurotransmitters, people tend to think more quickly, react more rapidly, feel more attentive and experience greater mental endurance and energy.

Wurtman's studies show that eating only three or four ounces of low-fat protein foods, either alone or with carbohydrates, will make tyrosine available to the brain if the brain is using up its current supply and needs to replace it. This process will have an energizing effect on the brain, increasing alertness, accuracy, and motivation. The brain will respond more quickly, prepared for mental challenges.

The best foods to choose for heightening brain power include three or four ounces of any one of the following:

fish
chicken without the skin
veal
very lean beef with all the fat trimmed
nonfat or low-fat yogurt
low-fat cottage cheese
tofu
lentils, dried peas, or beans

Avoid large meals and foods high in fat because they tend to dull the mind and slow mental processes. They produce lethargy and drowsiness. You must avoid foods such as lamb, pork and pork products, luncheon meats, hard cheeses, fatty beef, whole milk, regular yogurt (as opposed to nonfat or low-fat), butter, mayonnaise, fried foods, fatty meats, creamed soups, and rich gravies and sauces.

In addition to tyrosine, another amino acid, tryptophane, is found naturally in food and significantly affects our brain's ability to function. I am referring to tryptophane in its *natural* form, *not* in a pill. Naturally occurring tryptophane, which is found in carbohydrates, is the principal ingredient in the neurotransmitter serotonin.

Serotonin has a calming effect on the mind. It reduces feelings of tension and stress. Concentration and focus improve. Distractions are more easily eliminated. Reaction time may be slowed. And, if you eat serotonin-producing foods in the evening or before bedtime, they may induce drowsiness or sleep.

Eating only one or one-and-a-half ounces of carbohydrate alone, without protein, will increase your brain's level of serotonin. You will experience a calming, relaxing, more focused effect. Your powers of concentration will increase and your feelings of anxiety and frustration will ease. People who are 20% over their recommended weight, or premenstrual women may require two or two-and-a-half ounces of carbohydrates to get this effect.

The best foods to choose for producing a calming, focused effect on the brain include:

cereals	crackers	popcorn
muffins	pasta	potatoes
bagels	rice	bread
pancakes	barley	corn

Here is a summary of recommended health guidelines:

1. Make it a rule to get seven to eight hours of sleep regularly, especially before tests.

2. Allow time each day to meditate or relax. This is particularly important if you had little sleep or experienced insomnia the night before a test.

3. Participate in a program of moderate exercise for at least 20 minutes each day or every other day. Don't stop exercising around test time.

4. Eating for high mental energy requires that you eat small, low-caloric, low-fat meals and that you get proteins into your system before, or along with, carbohydrates. It is important not to begin the meal with carbohydrates. The first few bites of food should be of protein. Remember: You need only three or four ounces of protein. High amounts of protein may actually have a reverse effect on some people. Ideal foods include fish, chicken, veal, nonfat yogurt, tofu, lentils, and peas.

5. Time your high energy meal to occur about two hours before you must meet the mental challenge of taking a test.

6. To help you get adequate sleep before the test, eat high carbohydrate, low-protein (or even better, no protein) foods for dinner or in the evening to induce drowsiness and sleep.

7. If at any time you feel mentally frustrated, scattered, and can't seem to settle down, eat small amounts of carbohydrate to help calm you and focus your concentration. Although carbohydrates in the form of sugar also have this effect, be sure to eat no more than one or two ounces. Too much sugar causes negative emotional reactions in some people. Ideal carbohydrates to help settle you down are popcorn, bread, crackers, muffins, bagels, potatoes, rice, and pasta.

8. Don't overeat or eat foods with high fat content. These will only make you lethargic and dull.

9. Avoid excessive use of caffeinated beverages, such as coffee or soda. If you don't normally drink caffeine, don't start now, particularly not before a test.

PUTTING IT ALL TOGETHER

Use this Math Test Anxiety Reduction Checklist *before* your next exam.

EXERCISE 9-12: MATH TEST ANXIETY REDUCTION CHECKLIST

_____ I've reviewed and worked out lots of problems so I know my material out of context.

_____ I know the format and content of my upcoming math exam.

_____ I know how many questions will be on my exam and its duration.

_____ I've given myself several practice exams.

_____ On practice exams, I've noted areas of difficulty so I can strengthen them.

_____ I've analyzed my past pattern of typical errors so I can be alert to them on my exam.

_____ I've gotten seven to eight hours of sleep in the days prior to the exam.

_____ I've kept up a regular program of moderate exercise.

_____ I've practiced relaxation exercises along with positive visualization in the days and the half-hour before the exam.

_____ I've eaten a small meal of low-fat protein one to two hours before the exam and avoided too much caffeine.

_____ I'll arrive at the exam on time and avoid talking with others.

_____ Throughout the exam, I'll remain calm, relaxed, and positive, checking my breathing often.

_____ I will say positive self-statements to myself and push away all disturbing or distracting thoughts.

_____ I will write out all my formulas and key ideas on the top corner of my exam sheet before beginning the test.

_____ I'll quickly read through the exam, note point values, and schedule my time accordingly.

_____ I'll proceed comfortably throughout the exam, working first on the problems that come most easily to me.

_____ I'll carefully read the directions to all problems and circle significant words to avoid misinterpretation.

_____ After finishing the exam, I'll check my answers, proofread for omissions, and check for my typical errors.

_____ I'll leave and reward myself for a job well done!

SUMMARY

Don't allow test anxiety ever again to handicap your math achievement. What once was an overwhelming and fear-producing event can now be handled in a constructive, positive, growth-enhancing manner. This section of our road map focused on methods for markedly reducing math test anxiety. I've stressed good nutritional habits, adequate sleep, relaxation and exercise, and effective test preparation and test-taking strategies. In particular, I highlighted ways to deal with psychological factors through a special three-pronged approach. This approach includes a combination of shifts from negative to positive self-talk, calming techniques, and the utilization of my specially designed Math Test Anxiety Reduction Visualization. I concluded this section with an exercise to put it all together, the Math Test Anxiety Reduction Checklist.

CHAPTER

10

OPEN THE DOORS TO YOUR FUTURE

I was in college before I realized that mathematics is not cold. Mathematics is not like a piece of colorless cold marble—it has feeling, it has a warm temperature; mathematics has a beauty; it has a pulse; it has a heartbeat.

Bill Cosby
"Math: Who Needs It?"
PBS Video

If you think that math is just a bunch of numbers and formulas obscuring your appreciation of the world around you, you've got it backward. Math is humankind's constant effort to describe (and put to use) the mysterious workings of the universe. Lacking systems of measurement, primitive cultures could not produce machines, manufactured goods, or reproducible structures. Eruptions and earthquakes, storms, and celestial events were attributed to superstitious, and therefore nonverifiable, causes.

How times change—and so, too, do the needs and demands of an increasingly complex society. Seismic analysis, plate tectonics, meteorology, and astrophysics have created order out of seeming chaos. But our incessant need for precise measurement and manipulation of data has outstripped anything ever before imagined.

" STICK TO ANIMALS, THOG."

No longer just the language of science, mathematics now contributes in direct and fundamental ways to business, finance, health, and defense. For students, it opens doors to careers. For citizens, it enables informed decisions. More than ever before, Americans need to think for a living; more than ever before they need to think mathematically.

Everybody Counts: A Report on the Future of Mathematics Education
National Research Council

Math can open the doors to your future. To be an active, concerned member of this world, you must use the power of math. To be successful in school; have a rewarding, stimulating career; get the jobs you want; be an involved citizen; have a knowledge of personal finances, the nation's economy, and the technological advances of modern day society—all these require that you have an understanding of math. This final section of our road map explores the importance of math in choosing your career, in assuring higher salaries, and in every aspect of your life ranging from home and leisure activities to your civic and community life.

WHY IS MATH SO IMPORTANT?

Math is far more than the ability to calculate, memorize formulas, or solve equations. Many students don't understand this. Math trains your mind to think logically and succinctly. It requires you to perceive patterns, observe relationships, clarify and critically analyze problems, deduce consequences, formulate alternatives, test conjectures, estimate results, and enhance all of your problem-solving abilities. By sharpening your reasoning and thinking skills, you can become more productive in all aspects of your life.

Math provides you with the resources to comprehend the barrage of information that is communicated to you each day. It gives you the ability to be a more critical reader of anything you read, from newspaper reports to research articles to insurance policies and loan documents. Math logic, reasoning, and thinking ability help you ascertain possible risks or fallacies, to unearth biases, and to come up with suggestions and alternatives.

It's no wonder that so many careers require math skills. Employers want to hire individuals who can solve problems, who can think clearly on the job, and who can deal with new ideas, ambiguity, and change. We are constantly being flooded with technological advances, new scientific discoveries, new knowledge. Are you prepared for this challenge?

EXCITING CAREER CHOICES

Whether you realize it or not, an understanding of math and mathematical problem-solving and reasoning ability is necessary to enter and advance in almost all jobs and careers. The National Research Council (NRC) reports that 75% of all jobs require proficiency in basic algebra or geometry as a prerequisite for licensure or training. In addition, more than three-quarters of all the nation's university degree programs require more advanced math such as calculus, discrete math, statistics, or comparable mathematics. An advanced level of math competency is one of the essential elements for comprehending the mathematical basis for the sciences, engineering, technology, computer programming, and business. Moreover, almost any technical job in the future will require computer skills and these, in turn, are largely dependent on math skills or mathematical thought processes. Students who avoid taking math may permanently close the door on all of these career options.

I often ask students who are overcoming math fears to complete this incomplete sentence: "If I were better in math ______________________________________." These are just a few of the responses I've received:

I'd be an engineer.
I'd get a better job.
I'd be an R.N.
I'd get my B.A. degree.
I'd be a doctor.
I'd finally pass the civil service exam.
I'd become a scientist.
I'd be able to do better on the graduate record exam and get into graduate school.
I'd pass the advancement exam for my job.
I'd feel great about myself and my future.

You needn't let math hold you back anymore. I encourage you to continue to practice all the excellent strategies offered in this workbook and to follow your dreams. Don't close the doors on the exciting careers of today and the future. *You can succeed in math; you are a winner.*

Exercise 10-1 offers you a partial list of the many interesting and often high-paying careers that require an understanding of mathematical reasoning and problem solving. See if you can answer some of the questions I've posed.

EXERCISE 10-1: CAREER OPPORTUNITIES WITH MATH

Here are more than eighty rewarding careers that utilize math skills and thinking ability. Whereas you may already know about some of these career opportunities, others are probably quite unfamiliar to you. With the help of a career counselor, a college academic advisor, or career resource materials in a library, answer the following questions:

What kind of work do each of these people do?
How is math used in this career?
What math courses would help me succeed in this career?
What salary would I expect to earn in this field?

accountant
actuary
aircraft design engineer
airline navigator
archaeologist
aerospace engineer
agricutural economist
agronomist
architect
astronaut
astronomer
attorney
audio engineer
automotive engineer
banker
bank teller
biochemist
biologist
biomedical engineer
biometrician
bookkeeper
buyer
business administrator
cancer research scientist
cartographer
cryptographer
chemical engineer
chemist
civil engineer
commercial pilot
company president/CEO
computer consultant
computer engineer
computer music programmer
computer scientist
computer systems analyst
contractor
credit analyst
design engineer
economist
elementary school teacher
estimator
financial analyst
flight engineer
food scientist
geodesist
geologist
geophysicist
high school teacher
industrial engineer
industrial traffic manager
information analyst
insurance salesman
linguist
mathematician
marketing manager
mechanical engineer
meteorological technician
meteorologist
musicologist
nuclear scientist
nurse
oceanographer
optometrist
orthopedic surgeon
payroll specialist

physician
psychologist
psychometrician
purchasing agent
real estate broker
radiologist
research assistant
sales manager
school principal
secondary school teacher
software specialist
stockbroker
technical writer
weapons system analyst

ENJOY GREATER ACADEMIC SUCCESS

Members of the National Council of Teachers of Mathematics (NCTM) indicate that students who progress through mathematics are more likely to succeed in their education than those who avoid it. A recent study by the College Board has also found that math is the important monitor of academic success in college. Their findings show that math is "the great equalizer" between high-income students and those who are disadvantaged, of a minority, or have low income. The gap between these two groups in terms of their persistence and attendance in pursuit of a college degree essentially disappears when the latter group masters algebra and geometry in high school. ("Helping Students," 1991)

SCORE HIGHER ON EXAMS

I repeatedly hear from students how knowledge in math has helped or could help when taking important tests in their lives. Bonnie, one of my students, recently reported that she had applied for a job with a large department store's catalog department that required a preemployment assessment exam. When she went for the exam, she was extremely nervous, as were the other applicants waiting for the test to begin. There in the waiting room she practiced the deep abdominal breathing exercise (described in Chapter 3). She said that the exam had mainly math problems and because she had practiced relaxation techniques, she was able to remain calm, self-assured, and confident while answering all the math questions. Bonnie did well on the test and got the job. Bruce, a part-time student, recently returned to school with the sole purpose of improving his math skills. He said he wasn't able to advance any farther in the management track at a major industrial plant because he couldn't answer the many math questions on his company's internal promotion test.

The study of mathematics provides excellent preparation for advancement in many careers and jobs. Would you believe that, according to the Institute of Education's research report entitled "Standardized Test Scores of College Graduates," mathematics majors have ranked the highest in test performance for the Law School Admissions Test (LSAT) and the Graduate Management Admissions Test (GMAT)? High school students already know the important role of math for the Standard Achievement Test (SAT). But did you know that training in mathematics would also ensure you greater success on the Graduate Record Exam (GRE), an exam required by most universities for entrance into graduate programs?

EARN HIGHER SALARIES

Because math is essential in engineering, business, science, economics, and all technological fields, individuals who enter these disciplines with a good grasp of mathematical concepts and reasoning ability have a greater likelihood of succeeding and progressing up the career ladder. *USA Today* and other sources have reported that there is a positive correlation between people's future income and both the number of math courses they take and how well they do in those courses. A bachelor's degree in engineering, for example, usually requires several semesters of calculus and then a semester of differential equations. Of all the degrees a student can receive at the undergraduate level, engineering is an example of one of the highest-paying fields one can enter. Recent graduates of programs in electrical engineering are receiving job offers with starting salaries ranging from \$34,000 to \$43,000. Not bad for your first job out of school!

MATH FOR ALL PRACTICAL PURPOSES

Math is a constant in all aspects of our lives. A working knowledge of statistics, for example, is essential for understanding market surveys, election-poll results, clinical research findings, environmental studies, courtroom cases, stock market models, agricultural planning, weight-loss programs, meteorology, games of chance, sports results, traffic control, television programming, advertising campaigns, population growth predictions, and psychological studies, to name a few.

In this section, we'll explore the major areas of your life in which an understanding of mathematical concepts is essential. Can you add more to each of these areas?

IN OUR HOME LIFE

1. Reading computer manuals to learn how to use personal computers and software packages.
2. Setting a household budget and computing your taxes.
3. Planning your finances: investing; developing your investment portfolio; reading mutual fund and stock annual reports and prospectuses.
4. Redecorating your home: measuring square footage for floor coverings, window coverings, reupholstering, and buying paint.
5. Buying furniture, clothes, or household items.
6. Comparing mortgage, home improvement, and car loans.
7. Comparing insurance policies; calculating insurance payments.
8. Understanding maps and drawings drawn to scale.
9. Figuring out the amount of gas needed for a long automobile trip.
10. Understanding the effects of inflation on your income.
11. When buying food, calculating the best buy based on comparison of cost per ounce or per pound.
12.

13.
14.

IN HEALTH ISSUES

1. Understanding debates about AIDS testing.
2. Understanding environmental issues such as acid rain, the greenhouse effect, and waste management.
3. Figuring the amounts of various nutrients in foods.
4. Evaluating outrageous scientific claims; having the ability to discriminate between rational and reckless claims.
5. Measuring correct dosages of medicines; errors can be fatal.
6. Understanding research studies on placebo effects of drugs.
7.
8.
9.

IN CIVIC ISSUES

1. Understanding debates on the federal deficit, tax rate increases, nuclear disarmament.
2. Evaluating data appearing on population growth, the gross national product, unemployment projections, crime in the nation, consumer price index, and new construction costs.
3. Understanding political polls.
4. Determining energy consumption needs of the nation.
5.
6.
7.

IN LEISURE TIME ACTIVITIES

1. When traveling, having the ability to change U.S. dollars to foreign currencies.
2. Figuring scores in sports wagers.

3. Playing games of strategy, chance, and probability.
4. Figuring out puzzles.
5.
6.
7.

IN SCIENCE AND INDUSTRY

1. Math has left its mark on all aspects of industry and modern science. It gives us an understanding of the world we live in.
2. Math is the language of research and investigation.
3. Exploring changes in the tax codes, developing CAT and PET scanners, designing modern airliners and cruise ships, locating natural gas reserves and mineral deposits, decoding DNA, constructing a large dam; math is the one constant.
4. Designing automated systems and robotics.
5. Analyzing traffic patterns to plan for road construction.
6. Planning airline flight schedules and flight paths.
7. Analyzing data transmissions from satellites orbiting the earth.
8. Computing payroll and deductions, profit and loss statements.
9. In the space program—calculating escape velocities, trajectories, orbital data.
10.
11.
12.

IN ART, MUSIC, AND CULTURE

Architecture is akin to music in that both should be based on the symmetry of mathematics.

Frank Lloyd Wright

1. Mathematical concepts artists often use to create patterns and designs. (This is particularly true with graphic artists.

Famous artists such as Albrecht Durer, Leonardo da Vinci, and M. C. Escher relied on an understanding of math to create their masterpieces.)

2. Mathematical ideas used in the ancient Japanese art of origami—lines and points of symmetry, congruences, ratios, and proportions of shapes.
3. Necessity of mathematical concepts and measurements in architectural design.
4. Producing notes on a musical scale by taking the integral ratios of a string's length.
5. The design of Native American rugs—use of lines of symmetry, reflection, congruences, and accurate geometric proportions.
6.
7.
8.

GO FOR IT!

What you have to make them see—more than anything—is the future—which is mathematics.

Bill Cosby
"Math: Who Needs It?"

It's up to you to take action and jump into the wonderful world of math. You have the tools to succeed. Take it one positive step at a time, assertively move forward, and you will accomplish your goals.

EXERCISE 10-2: JUMP IN AND TAKE ACTION NOW

In the space below list five action steps you are willing to take in the next six months as a result of completing this workbook. Make sure to write down your target date for each step taken.

Action Steps **Target Date**

1.
2.
3.
4.
5.

SUMMARY

Math is the one constant in all sciences' efforts to understand, manipulate, and control natural phenomena. But beyond its necessity in science, mathematics is relevant in all aspects of our lives. This chapter explored the importance of math in training us to think clearly. It also showed us that math can open doors to better career opportunities, greater academic success, and higher scores on entrance and job placement exams. We also examined the applications of math in home life, health issues, civic issues, leisure time activities, science and industry, and in art, music, and culture.

Congratulations! You've come far. You have completed the major stepping stones along the road to math achievement. You've made a commitment to succeed. You are a winner!

REFERENCES

Coue, E. (1922). *Self-mastery through conscious auto-suggestion*. London: Allen & Unwin.

Dunn, R., and Dunn, K. (1979). *Student learning styles: Diagnosing and prescribing programs*. Reston, VA: National Association of Secondary School Principals.

Dunn, R., and Dunn, K. (1978). *Teaching students through their individual learning styles: A practical approach*. Englewood Cliffs, NJ: Prentice-Hall.

Ellis, A. (1974). *Guide to rational living*. North Hollywood, CA: Wilshire Book Co.

Gray, M. (1981, March-April). Mathematics, the military, and the mullahs. *Newsletter: Association for Women in Mathematics, 4*.

Helping students develop math power, (1991, April 18). *Guidepost*, p. 11.

Hanna, G. (1989). Mathematics achievement of girls and boys in eight grade: Results from twenty countries. *Educational Studies in Mathematics, 20*, 225–232.

"Math: Who Needs It?" with Jaime Escalante, aired 1991, PBS Video.

Magnesen, V. (1983). A review of findings from learning and memory retention studies. *Innovation Abstracts, V*, (25).

National Research Council. (1989). *Everybody counts: A report to the nation on the future of mathematics education*. Washington, DC: National Academy Press, p. 1.

Pauk, W. (1974). *How to study in college* (2nd ed.). Boston, MA: Houghton Mifflin, pp. 52, 55–57.

Richardson, A. (1969). *Mental imagery*. New York: Springer, pp. 56–57.

Richardson, F. C., and Suinn, R. M. (1972). The mathematics rating scale: Psychometric data. *Journal of Counseling Psychology, 19*, (6): 551–554.

Robbins, A. (1986). *Unlimited power—The new science of personal achievement*. New York: Simon and Schuster, p. 274.

Sell, L. (1978). Mathematics—A critical filter. *The Science Teacher, 45*, (2): 28–29.

Suinn, R. M. (1972). *A program of systematic desensitization for mathematics anxiety* (Cassette tapes). Colorado State University.

Suinn, R. M., Edie, C. A., Nicoleti, J., and Spinelli, P. R. (1972). The MARS, a measure of mathematics anxiety: Psychometric data. *Journal of Clinical Psychology, 28:* 373–375.

Suinn, R. M., and Richardson, F. C. (1971). Anxiety management training: Nonspecific behavior control. *Behavior Therapy, 2:* 498–519.

The Resource Manual for Counselor / Math Instructors. Math Anxiety, Math Avoidance, Reentry Mathematics. (1980). Prepared for the U.S. Dept. of

Education, the Fund for the Improvement of Postsecondary Education. Project Director: Sheila Tobias. Washington, DC: The Institute for the Study of Anxiety in Learning, The Washington School of Psychiatry, p. 49.

Wurtman, J. J. (1986). *Managing your mind and mood through food.* New York: Rawson.

SUGGESTED READINGS

Fanning P. (1988). *Visualization for Change.* Oakland, CA: New Harbinger Publications.

Jacobs, H. (1982). *Mathematics: A human endeavor.* San Francisco: W. H. Freeman.

Kogelman, S., and Warren, J. (1978). *Mind over math.* New York: McGraw-Hill.

Mitchell, C. (1991). *Math anxiety: What it is and what to do about it.* Dubuque, IA: Kendall/Hunt Publishing.

Nolting, P. (1988). *Winning at math.* Pompano Beach, FL: Academic Success Press.

Paulos, J. A. (1988). *Innumeracy—Mathematical illiteracy and its consequences.* New York: Hill and Wang.

Smith, R. M. (1991). *Mastering mathematics—How to be a great math student.* Belmont, CA: Wadsworth.

Tobias, S. (1978). *Overcoming math anxiety.* New York: Norton.

Tobias, S. (1987). *Succeed with math.* New York: College Entrance Examination Board.

Turner, N. D. (1971). *Mathematics and my career.* Washington, DC: National Council of Teachers of Mathematics.

Walter, T., and Siebert, A. (1990). *Student success—How to succeed in college and still have time for your friends.* Fort Worth, TX: Holt, Rinehart & Winston.

PUBLISHERS OF COLLEGE COURSE OUTLINE SERIES

AMSCO School Publications, Inc., 315 Hudson St., 5th fl., New York, NY, 10013-1085.

Barnes and Noble's Outline Series, 10 E. 53rd. St., New York, NY, 10022

Barron's Educational Series, Inc., 250 Wireless Blvd., Hauppauge, NY, 11788.

Schaum's Outline Series/College Division, Subdivision of McGraw-Hill, Inc., Princeton Road, Hightstown, NJ, 08520.